Population, Economic Growth and Agriculture in Less Developed Countries

In 1950 the world population was 2.5 billion; fifty years later there are over 6 billion people. This demographic explosion has essentially occurred in the developing areas of the world. The key to understanding many contemporary development problems that have arisen from this rapid growth is in understanding the relationships between population and the economy.

This book offers an analysis of such relationships, encompassing a review of the major positions in the academic debate. Nadia Cuffaro begins with Malthus and the population implications of the Solow model and then discusses the theoretical background of the less pessimistic ('revisionist') positions in the population debate. Part Two considers population growth in relation to agriculture and focuses on technical progress and institutional adjustments. It is argued that simple functionalism leads to a misrepresentation of the problems and that there are many possible adjustment failures linked to the research system, property rights on land and poverty that greatly affect the environmental resource base. Finally the author looks to the future to consider biotechnology and the challenges for world agriculture.

Population, Economic Growth and Agriculture in Less Developed Countries will serve as a useful introduction and reference tool for students, academics and all with an interest in the population debate and development economics.

Nadia Cuffaro is a Researcher at the University of Cassino, Italy. She has also worked as Economist at the United Nations Secretariat, New York and as a Consultant for the Food and Agricultural Organization of the United Nations, Rome.

Routledge Studies in Development Economics

1 Economic Development in the Middle East
Rodney Wilson

2 Monetary and Financial Policies in Developing Countries
Growth and stabilization
Akhtar Hossain and Anis Chowdhury

3 New Directions in Development Economics
Growth, environmental concerns and government in the 1990s
Edited by Mats Lundahl and Benno J. Ndulu

4 Financial Liberalization and Investment
Kanhaya L. Gupta and Robert Lensink

5 Liberalization in the Developing World
Institutional and Economic Changes in Latin America, Africa and Asia
Edited by Alex E. Fernández Jilberto and André Mommen

6 Financial Development and Economic Growth
Theory and experiences from developing countries
Edited by Niels Hermes and Robert Lensink

7 The South African Economy
Macroeconomic Prospects for the Medium Term
Finn Tarp and Peter Brixen

8 Public Sector Pay and Adjustment
Lessons from five countries
Edited by Christopher Colclough

9 Europe and Economic Reform in Africa
Structural adjustment and economic diplomacy
Obed O. Mailafia

10 Post-apartheid Southern Africa
Economic challenges and policies for the future
Edited by Lennart Petersson

11 Financial Integration and Development
Liberalization and reform in sub-Saharan Africa
Ernest Aryeetey and Machiko Nissanke

12 Regionalization and Globalization in the Modern World Economy
Perspectives on the Third World and transitional economies
Edited by Alex E. Fernández Jilberto and André Mommen

13 The African Economy
Policy, Institutions and the Future
Steve Kayizzi-Mugerwa

14 Recovery from Armed Conflict in Developing Countries
Edited by Geoff Harris

15 Small Enterprises and Economic Development
The Dynamics of Micro and Small Enterprises
Carl Liedholm and Donald C. Mead

16 The World Bank
New Agendas in a Changing World
Michelle Miller-Adams

17 Development Policy in the Twenty-First Century
Beyond the Post-Washington Consensus
Ben Fine, Costas Lapavitsas and Jonathan Pincus

18 State-Owned Enterprises in the Middle East and North Africa
Privatization, Performance and Reform
Edited by Merih Celasun

19 Finance and Trade in Developing Countries
Edited by Rohinton Medhora

20 Contemporary Issues in Development Economics
Edited by B.N. Ghosh

21 Mexico Beyond NAFTA
Edited by Martín Puchet Anyul and Lionello F. Punzo

22 Economies in Transition
A guide to China, Cuba, Mongolia, North Korea and Vietnam at the turn of the twenty-first century
Ian Jeffries

23 Population, Economic Growth and Agriculture in Less Developed Countries
Nadia Cuffaro

24 From Crisis to Growth in Africa?
Edited by Mats Lundahl

Population, Economic Growth and Agriculture in Less Developed Countries

Nadia Cuffaro

Routledge
Taylor & Francis Group
London and New York

First published 2001
by Routledge
11 New Fetter Lane, London EC4P 4EE

Simultaneously published in the USA and Canada
by Routledge
29 West 35th Street, New York, NY 10001

Routledge is an imprint of the Taylor & Francis Group

Typeset in Times by
Florence Production Ltd, Stoodleigh, Devon
Printed and bound in Great Britain by
Biddles Ltd, Guildford and King's Lynn

British Library Cataloguing in Publication Data
A catalogue record for this book is available from the British Library

Library of Congress Cataloging in Publication Data
Cuffaro, N. (Nadia)
Population, economic growth and agriculture in less developed countries/Nadia Cuffaro.
p. cm. – (Routledge studies in development economics; 23)
Includes bibliographical references and index.
1. Developing countries—Population. 2. Developing countries—Economic conditions. 3. Agriculture—Developing countries. 4. Economic development. 5. Neo-Malthusianism. I. Series.
HB884.C84 2001
304.6′09172′4—dc21 2001019968

ISBN 0–415–20290–6

Contents

Figures

Tables

Acknowledgements

This book was written on the basis of ideas and materials developed for a series of lectures I gave at Cambridge University between 1993 and 1996. I am particularly indebted to Michele De Benedictis, Mohan Rao, Fabrizio De Filippis, Giorgio Gagliani, Giorgia Giovannetti, Marcello Gorgoni, Carl Ipsen, Alain Marcoux, Marcello Messori, Luca Salvatici and Daniele Terlizzese, for comments and moral support.

Frank, Guido and my parents were as usual great allies for my work.

I would also like to thank the following for granting permission to reproduce material in this work:

Figure 1, p. 227 from J. Eatwell, M. Milgate and P. Newman (eds) *The New Palgrave: Economic Development*, 1989, London and Basingstoke: Macmillan. Reproduced with permission from Palgrave.

Figure 2, p. 1069 from M. Rosenzweig and O. Stark (eds) *Handbook of Population and Family Economics*, 1997, Amsterdam: Elsevier. Reproduced with permission from Elsevier Science.

Figure 1.4, p. 23 and Figure 1.11, p. 39 from Barro, R.J. and Sala-i-Martin, X. *Economic Growth*, 1995, New York: McGraw-Hill. Reproduced with permission from Robert Barro.

Figure 1 from Becker *et al.*, *Journal of Political Economy*, 98: 5 (1990): S17. Reproduced with permission from The University of Chicago Press.

Figure 4.2, p. 91 and Figure 5.6, p. 134 from Hayami, Y. and Ruttan, V. *Agricultural Development: An International Perspective* 1985, Baltimore: The Johns Hopkins University Press. Reproduced with permission from The Johns Hopkins University Press.

Figure 2.4, p. 29 from J.R. Anderson (ed.) *Agricultural Technology: Policy Issues for the International Community*, 1994, CAB International and the World Bank. Reproduced with permission from CAB International.

Figure 1, p. 1154 and Figure 2, p. 1155 from Cuffaro, N. *World Development*, 1997, 25: 7. Reproduced with permission from Elsevier Science.

Quotations from Bloch, M. (1971) 'The Rise of Dependent Cultivation and Seignorial Institutions', in M.M. Postan (ed.) *The Cambridge Economic History of Europe*, vol. I, Cambridge: Cambridge University Press. Included with permission from Cambridge University Press.

1 Introduction

This book intends to offer an analysis of the relationships between population, economic growth and agriculture in less developed countries – encompassing the discussion of the different positions in this debate – that could interest university students in development and population and their teachers and generally readers with a background in economics and an interest in the population debate.

A difficult task in a project like this is to select the topics to be included, and those that, although interesting and relevant, are less so in the perspective chosen and are therefore excluded. The perspective that has guided the selection here is the following.

Population pessimism is based on two major grounds: Malthusian concerns about the finiteness of natural resources and concerns about the (per capita) capital endowment of the economy. This book argues that population optimism, on the other hand, is very much related to the extension of a neoclassical perspective to the theory of institutions. Agriculture is both an important part of the development problem and the traditional (Malthusian) basis for population pessimism. Agriculture is also a sector where the more recent, less pessimistic ('revisionist') position has found relatively solid empirical and theoretical support. By and large the theoretical background of the analyses that suggest population optimism for agriculture lies in a more general belief that the power of the invisible hand works also for the institutions. However, it is argued here, both in general and in the case of agriculture in less developed countries, that simple functionalism leads to a misrepresentation of the problems which arise with population growth. There are many possible failures of adjustment to population, which usually have consequences on the environmental resource base.

Given this perspective, the book illustrates the models at the basis of population pessimism and the theoretical perspectives suggesting population optimism in Part I and the population–agriculture links in Part II.

Part I focuses on theory more than on empirical evidence, and also serves the purpose of putting the population–agricultural development analysis in a general theoretical perspective. Chapter 2 ('Population pessimism') presents the Malthus model and the Solow growth model. The former represents one polar, pessimistic view of the population–development linkages and it is a necessary point of departure for any discussion of the relationship between demographic and economic variables. The latter has represented economic orthodoxy in growth theory for a long time, is the reference model for all modern neoclassical growth theory and is included in this chapter because its population implications too are pessimistic. Chapter 3 outlines the theoretical background of a more recent and less pessimistic ('revisionist') view of the population–economic development links. It is a selective review encompassing the contribution of new institutional economics – and especially North and Thomas' (1971, 1973) neoinstitutionalist view of the impact of population on modern European growth; a critique to the 'tragedy of the commons' notion, in reference to local commons; the population and savings links and, finally, the role of population in recent neoclassical growth theory.

Technical progress is the first topic of Part II – the importance attributed to technological progress within neoclassical growth theory is paralleled by the importance of the theories of innovation for agriculture. The ideas discussed in Chapter 4 are Boserup's (1965) notion of population-induced technical progress in agriculture and the Hayami and Ruttan induced innovations model (1985). The responses of institutions (including property rights on land and the institutional factors influencing the extension and quality of the irrigation infrastructure) – are crucial for the adjustment mechanisms foreseen by both models. These topics are therefore taken up in Chapter 5. Chapter 6 examines the possibility of adjustment failures linked to the research system, property rights on land and poverty; Chapter 7 discusses the environmental implications of such failures, focusing on population-induced land degradation and Chapter 8 concludes the book with a discussion of the future research challenges in agriculture, including the 'promises' of biotechnology.

Intrahousehold issues and the links between population and natural resources other than land are only marginally mentioned, the former in the context of the discussion of property rights on land, the latter – water and forests – only in terms of their relation with agriculture.

In the case of forests, one may however observe that there is a link between agricultural intensification and deforestation: if more can be produced on existing land there is less incentive to use the extensive margin – i.e. to clear forests for cultivation – and communities may also be less dependent on forest products. Hence, our topic – the capability to intensify agricultural production in response to population growth – is relevant for the relationship between demographic variables and deforestation.

The neoclassical approach – both to growth theory and to institutions – holds a large space in this presentation, because it has been the predominant view underlying many aspects of the population debate. In particular, population optimism – as opposed to Malthusian concerns over diminishing returns and to the 'resource shallowing' effect of the neoclassical growth model, is essentially based on the idea that population growth, by changing certain relative prices, will produce responses – chiefly in terms of property rights and technical progress – that essentially take care both of Malthusian and of traditional neoclassical concerns.

However the analysis in this book exposes some strong limitations of this view, in the first place in relation to the issue of property rights on land. Property rights on land are the most important population-related institution in agriculture (and in general the most important institution for this sector). Yet, all the property rights paradigm tells us about adjustments in response to population growth is that individual private property will eventually emerge. As for the pattern of land ownership, since there is abundant evidence that land concentration leads to inefficiencies, on the whole we should observe that such concentration does not emerge, or at least does not persist. This of course is far from being close to the historical evidence. Obviously there are important missing elements in a paradigm that explains institutions essentially as efficient outcomes of the interaction between individuals.

Some notion of power must be introduced in the reasoning if one is to explain land concentration. Since land concentration is strongly related to poverty and landlessness and these in turn affect people's ability to acquire food, the topic is very relevant for our subject. Furthermore, large inequalities tend to reduce the probability of success of collective action in irrigation and in the management of local commons, to the extent that the appropriation of a large share of resources on the part of some members of the community increases the pressure on the residual commons.

Second, functionalist analyses tend to ignore the problem of the timing of adjustments, which, with fast population growth, is a critical

Table 1.1 World population growth

Demographic index	*10000* BC	*0*	*1750*	*1950*	*2000*
Population (millions)	6	252	771	2530	6235
Annual growth (%)	0.008	0.037	0.064	0.569	1.812
Doubling time (years)	8369	1854	1083	116	38

Source: Livi Bacci (1997).

issue. The functionalist bias of neoclassical neoinstitutionalism has been subject to many criticisms, which are outlined in Part I. The institutional analyses usually recognise that many factors may lead to suboptimal outcomes and/or slow down the adjustment pace. But even in the best of all possible worlds, population-induced adjustments may lag behind the type of fast demographic growth experienced by contemporary less developed countries.

Table 1.1 gives a broad idea of the acceleration of the pace of population increase that has occurred world-wide over human history. The acceleration following the industrial revolution is very great, but that which has occurred since 1950 translates into 'numbers' that are incomparably higher. The size of population by continent since 1750 and projections for 2050 are visualised in Figure 1.1.

Although the world's growth rate of population reached a peak in the late 1960s and has declined since, annual population additions have continued to rise because such growth rates are applied to rapidly expanding totals.

The difficulty of rapid adjustments in the institutional sphere – where slowly changing cultural factors have a greater role – usually has consequences in terms of resource depletion. This is an outcome which is mostly illustrated in this book with examples from agriculture. Poverty, difficult natural environments and inequality in the structure of asset ownership make adjustments more difficult and interact negatively with the 'time' factor.

Finally, the adjustments induced by relative market price changes will not address the problems of externalities, such as those arising from the intensification of agriculture based on the massive use of chemicals.

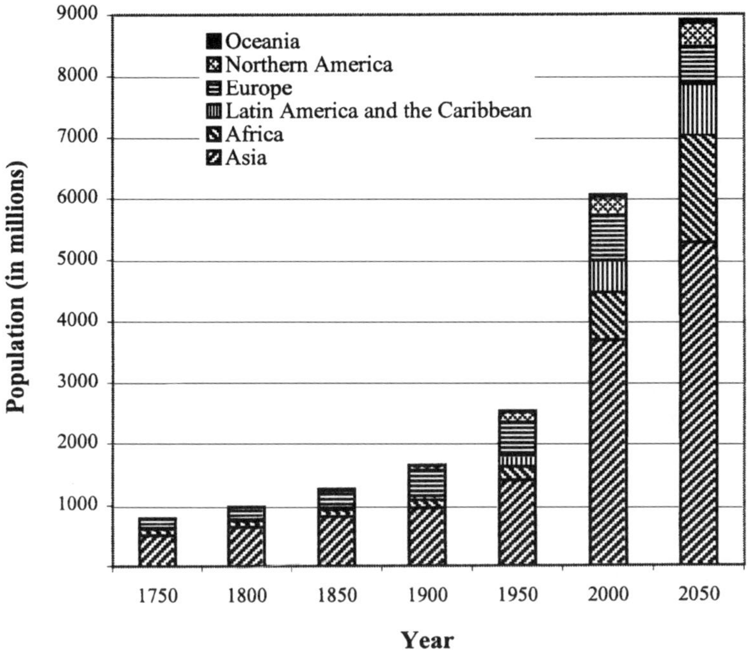

Figure 1.1 Population of the world and its major areas, 1750–2050.
Source: United Nations (1999a, b)

In the case of agriculture, on one hand it appears that population growth induces adjustments that on the whole make it possible to meet growing food needs and to maintain agricultural incomes (although much less successfully) and land quality; on the other hand there have been many local failures involving land degradation – or population-related cycles of land degradation – and agricultural intensification has often occurred with considerable negative environmental effects.

Part I

Population growth and economic development

2 Population pessimism

Malthus's theory of population

The model

Malthus's *An Essay on the Principle of Population* was first published in 1798 and then in six successive editions. The starting point of the analysis is the recognition of the basic difference between the laws that regulate population growth and the laws that regulate food production growth. In theory, Malthus maintains, there are no limits to what he defines the power of population.

> A thousand millions are just as easily doubled every twenty-five years by the power of a population as a thousand. But the food to support the increase from the greater number will by no means be obtained with the same facility . . . When acre has been added to acre till all the fertile land is occupied, the yearly increase of food must depend upon the amelioration of the land already in possession. This is a stream which, from the nature of all soils, instead of increasing, must be gradually diminishing.
>
> (Malthus 1826: 10)

The hypothesis here is that there are decreasing returns to labour with a fixed land endowment, and this generates a conflict between population and its means of subsistence: indeed, the most quoted paragraph of the Essay predicts that

> population when unchecked goes on doubling itself every 25 years, or increases in a geometrical ratio . . . the means of subsistence, under circumstances the most favourable to human industry, could not possibly be made to increase faster than in an arithmetical ratio.
>
> (Malthus 1826: 10–12)

There are however 'checks' to population, which are divided into two general categories: preventive checks and positive checks. The latter

> include every cause, whether arising from vice or misery, which in any degree contributes to shorten the natural duration of human life. Under this head therefore may be enumerated all unwholesome occupations, severe labour and exposure to the seasons, extreme poverty . . . the whole train of common diseases and epidemics, wars, pestilence, plagues and famines.
>
> (Malthus 1826: 16)

The former arise either from 'moral restraint' within marriage, or from all other ways in which sex may be separated from procreation and which, '. . . clearly come under the head of vice' (Malthus 1826: 16). Malthus maintains however that in spite of such checks

> there are few states in which there is not a constant effort in the population to increase beyond the means of subsistence. This constant effort as constantly tends to subject the lower classes of society to distress, and to prevent any great permanent amelioration of their conditions.
>
> (Malthus 1826: 17)

The dynamics of the interaction between population and its means of subsistence is described starting from a situation of satisfactory equilibrium. Population then tends to increase faster than its means of subsistence. This implies a decrease in the price of labour relative to the price of food, which in turn translates into both preventive and positive checks to population and into a more intensive use of land, until the original equilibrium between population and food is restored. The mechanism just described will then restart, but it will always result, at end, in a population that lives at subsistence level. This oscillation, according to Malthus, has not emerged as it should have from historical records because, to use his words 'the histories of mankind which we possess are, in general, histories only of the higher classes' and because of the failure to observe the correct variable, that is the real, rather than nominal, price of labour.

We therefore have a model in which food availability is a core argument of the population scare. Population is endogenous (the birth and death rates vary in response to changes in the standard of living), and there are decreasing returns to labour with fixed land endowment.

When population grows, since food production growth is less than proportional, real wages decrease and, as a consequence, the birth rate declines and the death rate increases, until the original equilibrium is restored: food production therefore establishes the limits to population growth.

This model can be graphically described as in Figure 2.1 (Weir 1988). Figure 2.1(a) shows a relation between real wage (or income per capita) and the size of population reflecting the hypothesis of decreasing returns; 2.1(b) describes demographic behaviour: when the real wage declines mortality increases (the positive check) and fertility declines (the preventive check). Point *e*, with zero growth of population and a given standard of living, is a stable equilibrium, in the sense that if either population or the real wage assume a value different from w^* and p^* this would set in motion a mechanism which tends to restore the original equilibrium. For example, a larger population would reduce the real wage, and cause a decline in the birth rate and an increase in the death rate, bringing the system back towards *e*.

Population is endogenous in a sense that is close to the one found in biological models of animal populations, where food supplies or predator population limit the size of the population. Malthus – and the classical economists in general – viewed population dynamics – its fertility and mortality – as elements of a system that includes the size of the population in relation to the means of subsistence.

> In this system we look for feedback between population levels and population change. This Classical approach is quite remote from modern demographic theory of population change, which focuses on detailed analysis of each vital rate in isolation from population

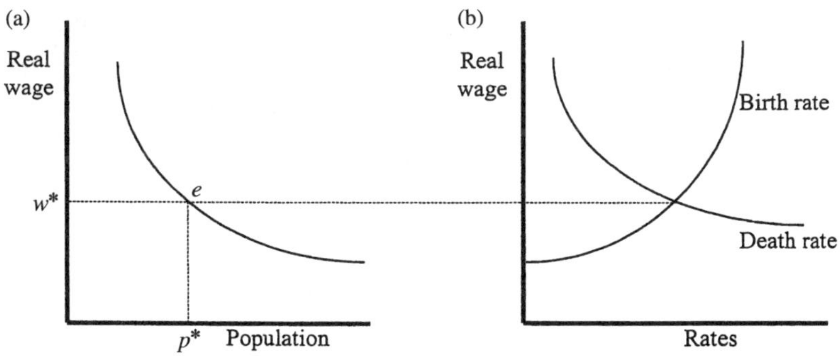

Figure 2.1 The Malthus model.

Source: Weir (1988)

> size. However, it is the way many anthropologists view prehistoric population dynamics, the way many social historians view historical population change, and the way biologists view animal populations. It lies behind the prescriptions of ecologists and environmentalists for contemporary population limitation.
>
> (Lee 1997: 1064–1065)

This approach is very different from the one adopted in the recent literature on the economics of the family, discussed in the next chapter. In models belonging to this strand of theory population is endogenous because families are supposed to make conscious fertility choices, mostly in a utility-maximising framework, in response to changes in economic variables.

The historical evidence

Historical examinations of the data on population, wages and prices for several European countries before Malthus' time tend to accord with the notion that the cyclical succession of population growth and decline is explained by the long-term modifications of mortality and nuptiality in reaction to periods of improving or worsening living standards.[1]

Indeed some of the most influential interpretations of the long-term socio-economic trends in medieval Europe are based on the cyclical dynamics of the Malthusian model. Since once population grows its economic consequences – diminishing returns, falling wages, rising food prices and increasing rents – set a self-correction mechanism, population then must necessarily decline again, starting a new cycle. This centuries-long two-phase movement has been seen as representing the very essence of 'traditional economy' and neo-Malthusianism is the predominant orthodoxy in the economic history of preindustrial societies. In the words of Emmanuel Le Roy Ladurie 'From the fourteenth to the seventeenth centuries included the economy is servant rather then master, led rather than leading . . . in the last analysis it is meek enough before the great forces of life and death. And, as for politics or the class struggle, their moment of power is still to come.'[2]

Demographic growth characterises both the twelfth and the thirteenth centuries. That growth was associated with low wages and rising food prices, and therefore with the decline of the living standards of the peasants. Both the widespread famine of 1315–1317 and the plague – which spread over Europe between 1347 and 1351 and became endemic thereafter, resulting in a cycle of population decline which appears to have continued for a century – have been interpreted by many historians as Malthusian checks.[3]

After the year 1300 different parts of Europe experienced bad agricultural years. Poor harvest, in the conditions of medieval Europe, would most easily translate into famine for two reasons: because the monoculture of corn was widespread and hence there was hardly an alternative in case of harvest failure; because local shortages could not be easily filled through trade, given the conditions of transport and the insufficient development of markets.[4] The endemic undernourishment that afflicted Europeans during the fourteenth century contributed to high mortality directly and indirectly, as a weaker population was more defenceless against epidemics. Indeed, the first great explosion of the plague in several Northern European regions was preceded by years of bread shortage. The plague first hit Italy in 1347 and spread through most of Europe within two years, reducing the population on the average by 30 per cent, but in some areas by as much as 70 per cent. After this great wave, the epidemic resurfaced periodically in different areas and in some remained endemic.

The demographic catastrophe of the last two centuries of the Middle Ages was accompanied by the growth of the average size of land holdings everywhere in Europe, by a declining trend of the absolute and relative price of corn, by a general, and mostly substantial, growth of all categories of wages and by a reduction of the value of arable land. This drastic reversal of the man/land ratio, and hence of relative prices and living standards set a new cycle of population growth, in the last half of the fifteenth century.

Figure 2.2, relating population and real wages in Europe suggests an inverse association between the two variables: population increased sharply during the thirteenth century and high population levels at the beginning of the fourteenth century were associated with low wages. Population declined and wages increased sharply following the Black Death of 1348 and subsequent epidemics. From the middle of the fifteenth century population increased again and real wages declined.

It must be noted that the Malthusian interpretation covers the full spectrum of socio-economic trends in medieval society, as the whole structure of customary rights and obligations that defined the landholding relationship – including those provisions that defined the free or unfree status of the peasant – are supposed to fluctuate with the population pressure on land. Hence, in this view, the disappearing of serfdom would be a consequence of the demographic decline of the fourteenth and fifteenth centuries.

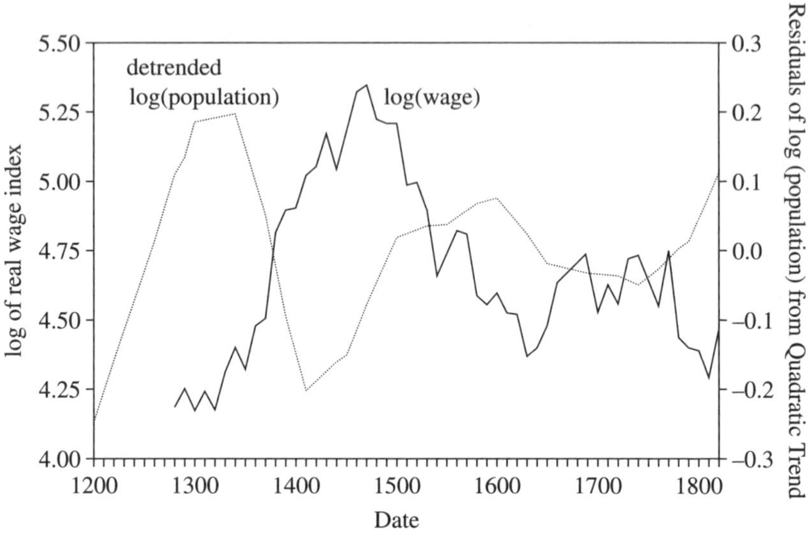

Figure 2.2 Population and real wages in Europe, 1200–1830.
Source: Lee (1997)

Postan's (1971) account of the history of medieval agrarian society in England is an authoritative example of this approach. The rising population trend between the eleventh and fourteenth centuries resulted in a large accumulation of paupers at the bottom of the social scale. Postan describes thirteenth-century England as a country densely occupied in conditions of very low land productivity, and compares it to the densely populated under-developed countries of the modern world. With growing density, the actual payments for land tended to rise, as landlords increasingly used a variety of devices – such as supplementary taxes and lump sum payments from incoming tenants – to circumvent the low, and in principle fixed, customary rents.

The Malthusian checks would then follow, since

> A population whose land hunger was so acute and whose proportions of petty holders were so large, must have been moving ever nearer to the very margin of subsistence. Hence the high death rates which most invariably followed bad harvests. The unprecedented heavy death toll which these harvest failures took, like the

> toll which similar calamities have taken in certain countries in modern times, could not have been due to natural causes alone, but to the calamity-sensitive constitution of society, and above all, to the precarious balance between men's needs and the productivity of their holdings characteristic of an over-populated country and an overextended agriculture.
>
> (Postan 1971: 564–565)

Postan holds that the Black Death of 1348–1349, although it carried off at least a third of the total population – by itself could not have been sufficient to keep the population back for more than a century – since the first indications of demographic recovery begin to appear only in the second half of the fifteenth century. Hence the demographic decline in the later Middle Ages is seen as the result of a combination of the effects of high population pressure on land and the plague.

The Malthusian checks and demographic decline are associated with a movement of increasing land to labour ratio, a contraction of cultivated land, decreasing rents and rising wages at the beginning of the fifteenth century. This in turn, according to Postan, fostered a major institutional development: the decline of serfdom. Indeed, in most parts of England the landlords, in order to retain their tenants and to let out their vacant lands, proceeded to lighten the burdens of payments and to cancel altogether the remaining labour dues.[5]

The sixteenth century witnessed the next cycle of population expansion, accompanied again by rising agricultural prices and falling real wages. However, the Malthusian reaction of the next century was to be both less severe and much less uniform than that which had occurred in the previous cycle. The explanatory power of the population model seems to fade from the sixteenth century. Between 1500 and 1750 Europe experienced dramatically contrasting trends of economic development: continuing long-run stagnation accompanying population increase in some areas – as in France – the spectacular emergence of an entirely new pattern of relatively self-sustaining growth accompanying increases in population in other areas, as in England. 'By the end of the seventeenth century English population had returned to its high, late thirteenth-century levels, but there was nothing like the demographic pattern of seventeenth-century France, no phase B following inescapably from phase A. Instead we have the final disruption of the Malthusian pattern and the introduction of a strikingly novel form of development' (Brenner 1976: 42).

'Malthusian' years: the 1960s and 1970s

In a review of the debate on the economic consequences of population change in the Third World, Kelley (1988a: 1698) observes that the 1960s and 1970s witnessed a strong emphasis on the negative, supply side impacts of population and that Malthusian concerns regarding the population–natural resources balance had never achieved greater popularity since the early nineteenth century.

This development in the perception of the relation between population and the economy is viewed as the result of several factors. Some of these were factual developments, namely, the unprecedented rates of population growth in the Third World and the promotion and implementation of birth control policies and programmes by governmental agencies.

But the population pessimism arising from different strands of economic analysis also played an important role. Coale and Hoover (1958) produced a very influential study of the impact of population on investments and savings; the neoclassical growth model, which was to exercise a long-lasting influence on economists' thinking about growth, had pessimistic population implications and as Birdsall (1988) remarks 'the emphasis in the general development literature was on lack of capital (and savings) coupled with surplus labour in agriculture as the major constraints to economic growth'.

Furthermore the studies by J. Forrester (1971) and Meadows *et al.* (1972) applied a straightforward extension of the Malthusian notion of diminishing returns to many renewable and non-renewable resources. These studies, although severely criticised by economists – who contended that under a number of assumptions about technical change and production technology society could substitute man-made for natural capital[6] – received much public attention and greatly contributed to the popularity of Malthusian pessimism in the 1970s.

In a review of the debate among demographers, Hodgson (1988) describes the 'orthodoxy' which emerged in American demography after the Second World War as a perspective which viewed demographic trends as determinants of economic trends, rapid population growth as a cause of underdevelopment and lowering fertility as a means of facilitating development.

This did not accord well with the scientific consensus that had been reached by the mid-1940s on transition theory. The term 'demographic transition' indicates a broad pattern of secular change in birth and death rates which demographers had identified on the basis of the historical experience of presently developed countries. The starting

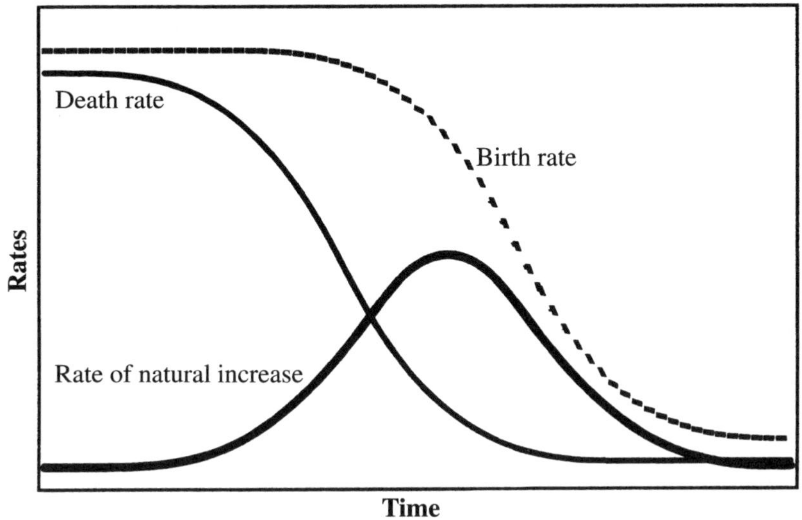

Figure 2.3 The demographic transition model.

point is a situation characterised by high birth and death rates, and nearly stationary population; in a next stage the death rate declines, because of improved standards of living and control of epidemics and other diseases, while fertility remains high (or decreases slightly), resulting in population growth. Next, the decline of mortality reaches a minimum while fertility declines, until a new equilibrium is reached, with low fertility and mortality rates and nearly stationary population.

In this framework demographic trends, and especially fertility decline, are seen as a consequence – and not as determinant – of structural change. The reasons for a reversal of perspective are, according to Hodgson, of three different types. One set of reasons is 'external' and includes the adoption of orthodoxy as a development strategy in the contest of the Cold War and the availability of funds for population control. Another set is of a scientific nature: transition theory had lost part of its explanatory power both in Western societies – where the 'baby boom' reversed the long-term decline in fertility – and in the less developed countries (LDCs) – where mortality decline had become divorced from the process of economic development.

In the LDCs mortality levels remained high until recently but declined very rapidly since the Second World War, chiefly as a

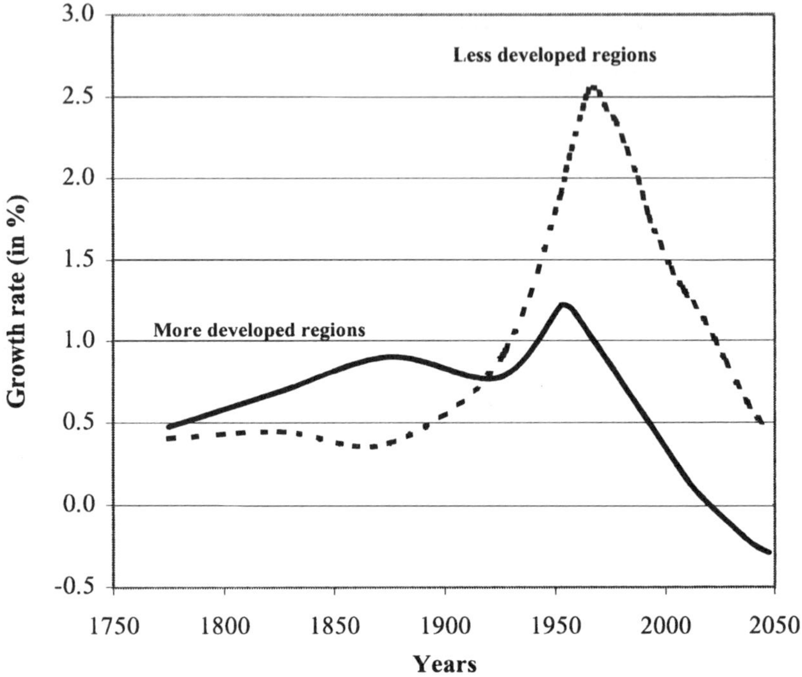

Figure 2.4 Population growth rates.

Source: United Nations (1999a, b)

consequence of public health measures, such as vaccinations, which were made possible by the transfer of scientific knowledge from the developed countries. Hence, while in the latter the reduction in mortality had occurred slowly, in parallel with the slow accumulation of health improving knowledge, in the LDCs the pace of such decline was much faster. Fertility behaviour on the other hand is linked to socio-economic and cultural factors that are likely to change slowly. Hence fertility decline lagged substantially behind, resulting in unprecedented rates of population growth in the LDCs: by the late 1960s the population growth rate had reached a peak of an average 2.5 per cent per year, a rate much higher than those observed historically in Europe at the same stage of 'demographic transition' (Bongaarts 1995; Livi Bacci 1997). The difference between the two paths of transition is illustrated by Figure 2.4.

Finally a role was played by some economists' views of the links between population growth and the economy.

The population implications of the neoclassical growth model are discussed in the next section; as for the Coale and Hoover (1958) analysis, it starts from the assumption that the main obstacle to economic growth in less developed countries is a deficiency of supply of capital.[7] The arguments that had the strongest influence on contemporary thinking about the population–development nexus regard the impact of the population age structure on savings and on investment composition. On the first point, Coale and Hoover observe that high fertility implies a high proportion of non-active population. In the case of India the growth of the labour force was less than proportional to population growth for two reasons: women's participation to the labour force was hampered by frequent births and by the larger numbers of children in the family; manpower use was not limited by the rate of growth of population, but rather by other factors. The high proportion of dependants, the 'burden of dependency' would have two negative effects on the economy: it reduces the rate of savings and it determines a diversion of investments from directly productive uses to programmes such as schooling and other forms of social expenditure.

Hodgson (1988) stresses the fact that demographers were well aware of the changing age structure of populations in LDCs, which resulted in a very high ratio of young dependants to the labour force, and this may have contributed to the impact that the 'burden of dependency' argument had on their views.

The empirical validity of the Coale and Hoover study has been largely questioned by successive analysis; however, its influence on the contemporary debate was so deep that their thesis 'eventually provided justification for birth control as a part of US foreign policy' (Kelley 1988a: 1699).

The population implications of the neoclassical growth model

The neoclassical growth model (Solow 1956, 1987) has exercised a decisive influence on the debate concerning the linkages between population growth and economic development. Although such debate has seen a variety of positions over time (and from it several specialised fields have developed) there is a unifying feature among many studies: a common reference to the neoclassical growth model and its conclusion that rapid population growth, by lowering the capital–labour ratio (an effect often referred to as 'resource shallowing' or 'capital widening') reduces the level (but not the growth rate) of per capita income. Such influence is not surprising, since the

Solow model has dominated neoclassical economists' thinking about growth for a long time.

The model can be represented through Equations (2.1) to (2.4).[8] Output (Y) is a function of capital (K) and labour (L) and the production function is neoclassical, i.e. it has three characteristics: it exhibits positive and diminishing marginal products with respect to each input and constant returns to scale to both inputs;[9] the marginal product of each input approaches infinity as the input goes to 0 and approaches 0 as the input goes to infinity (these last two properties are called *Inada conditions*). Savings (S) in the model is a constant fraction (s) of income – the saving rate s is given exogenously; investment (I) is equal to savings and the labour force grows at the exogenous constant rate n.

$$Y = F(K, L) \tag{2.1}$$

$$S = sY \tag{2.2}$$

$$I = S \tag{2.3}$$

$$L = L_0 e^{nt} \tag{2.4}$$

Since Equation (2.1) exhibits constant returns to scale it can be rewritten as

$$Y = F(K, L) = LF(K/L, 1) \tag{2.5}$$

Therefore the production function can be expressed in the 'intensive form' and be represented in a two dimensional space

$$y = f(k) \tag{2.6}$$

where $y = Y/L$ and $k = K/L$ i.e. output per worker is expressed as a function of capital per worker. If we assume that capital depreciates at the constant rate $\delta > 0$ the net increase in the stock of capital $\dot{K}$ (a dot over a variable denotes differentiation with respect to time) at a point in time equals gross investment minus depreciation

$$\dot{K} = I - \delta K = sF(K, L) - \delta K \tag{2.7}$$

Dividing both sides of this equation by L

$$\frac{\dot{K}}{L} = sf(k) - \delta k \tag{2.8}$$

The change in k over time $\dot{k}$ is[10]

$$\dot{k} = \frac{\mathrm{d}(K/L)}{\mathrm{d}t} = \frac{\dot{K}}{L} - n\left(\frac{K}{L}\right) = \frac{\dot{K}}{L} - nk \tag{2.9}$$

which, considering Equation (2.8) can be rewritten as

$$\dot{k} = sf(k) - (n + \delta)k \tag{2.10}$$

The first term on the right-hand side of Equation (2.10) is savings per worker, which in the model equates to investment per worker. The second term represents the investment necessary to equip new entrants in the labour force with the same average quantity of capital, and to substitute for capital that wears out. The $(n+\delta)$ term could also be interpreted as the rate of depreciation of k, as it tends to diminish the accumulation of the capital stock. The rate of growth of the capital/labour ratio is given by the difference between savings (and investment) per worker and what is required to keep the ratio constant as the labour force grows and capital depreciates.

Equation (2.10) is a differential equation in k whose graphical solution is depicted in Figure 2.5. The shape of the $f(k)$ and $sf(k)$ functions reflect the characteristics of the neoclassical production function. The $f(k)$ curve represents the relation between capital per worker and product per worker, and exhibits diminishing marginal productivity of capital; $sf(k)$ is the fraction of product which is saved (and invested); the ray from the origin represents the $(n+\delta)k$ function. Where the two curves intersect $(n+\delta)k = sf(k)$ and therefore $\dot{k} = 0$. At this point investment is just sufficient to maintain the capital endowment k for the new workers and to compensate for capital depreciation. If the economy starts out at a level of k lower than k^*, since at this level gross investment exceeds depreciation ($sf(k) > (n+\delta)k$), k grows. Thus the economy moves towards k^*. At that point, since savings will be just sufficient to cover depreciation, k does not change further. Similarly, if the economy starts with a level of k higher than k^*, $sf(k) < (n+\delta)k$, net investment is negative and the economy moves towards k^*. k^* is the long-run 'steady state' equilibrium.

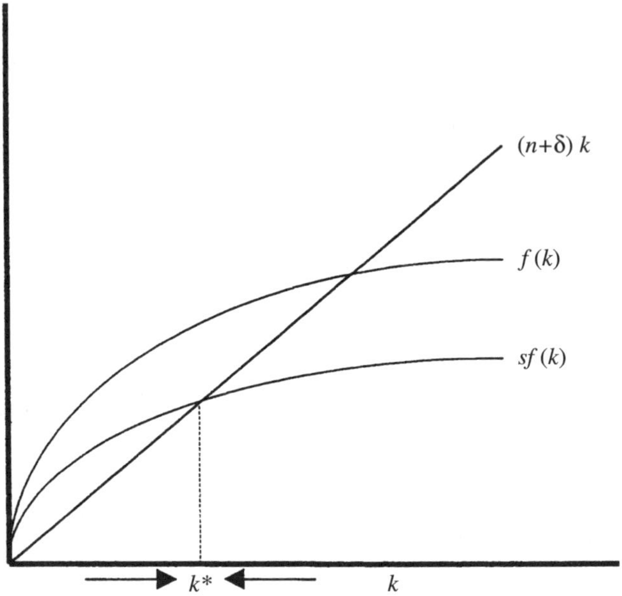

Figure 2.5 The dynamics of the Solow model.

The steady state is defined as a situation in which the per capita variables grow at a constant rate, which in this version of the Solow model is zero. The constancy of the capital–labour ratio implies that the capital stock expands at the same rate as the labour force. Since the model assumes constant returns to scale in the two inputs, output will also grow at the same rate and output per worker will be constant.

Another way of representing the dynamics of the model is to consider that, from Equation (2.10) the growth rate of k is

$$\frac{\dot{k}}{k} = \frac{sf(k)}{k} - (n + \delta) \tag{2.11}$$

Because of diminishing returns to capital $f(k)/k$ is relatively high when k is low, hence the first term on the right hand side of Equation (2.11) is a downward-sloping curve, which asymptotes to infinity at $k = 0$ and approaches 0 as k tends to infinity[11] (Figure 2.6). Therefore the steady-state capital–labour ratio k^* exists and is unique. Since k grows to the left of k^* and declines to the right of k^* the economy tends asymptotically toward the steady state k^*.

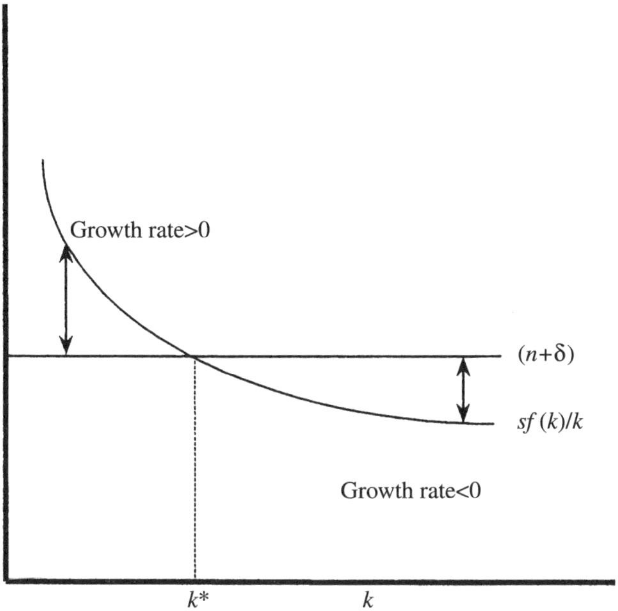

Figure 2.6 Dynamics of the Solow model: the growth rate of *k*.
Source: Barro and Sala-i-Martin (1995)

A change in one of the exogenous variables determines a change in the level of *k*. In Figure 2.7 a better technology or a higher saving rate moves the $sf(k)$ curve to the right. The growth rate of *k* becomes temporarily positive, but declines towards 0 as the economy approaches a new steady state in k^{**} .

These level effects can also be illustrated in terms of Figures 2.8 and 2.9.

A lower level of *s* shifts the $sf(k)$ curve downwards. Given the capital–labour ratio k^*, saving per worker is lower than the depreciation of *k*. The capital–labour ratio will therefore decline to a new steady-state level k^{**}.

A similar mechanism applies to changes in the growth rate of population and the labour force.

In Figure 2.9 a higher level of *n* rotates the $(n+\delta)k$ function. Investment becomes lower than 'depreciation', net investment is negative and the economy moves towards a new equilibrium in k^{**}, where both the capital–labour ratio and product per worker are lower. The

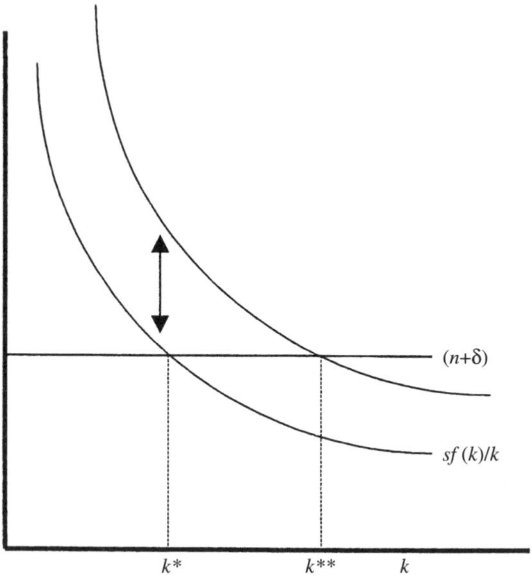

Figure 2.7 Effects from an increase in the saving rate or an improvement in technology.

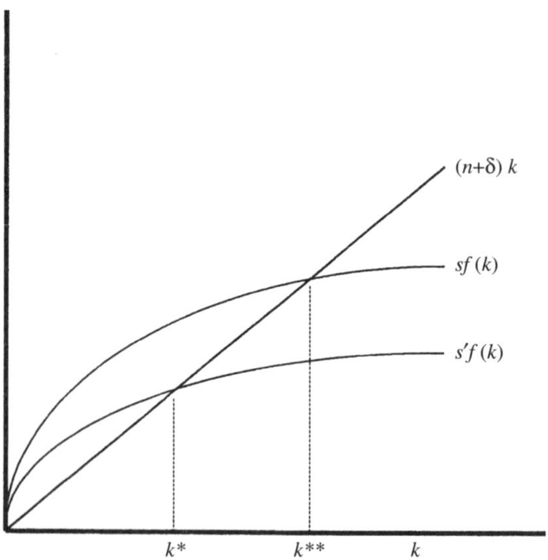

Figure 2.8 Effects from a decrease in the saving rate.

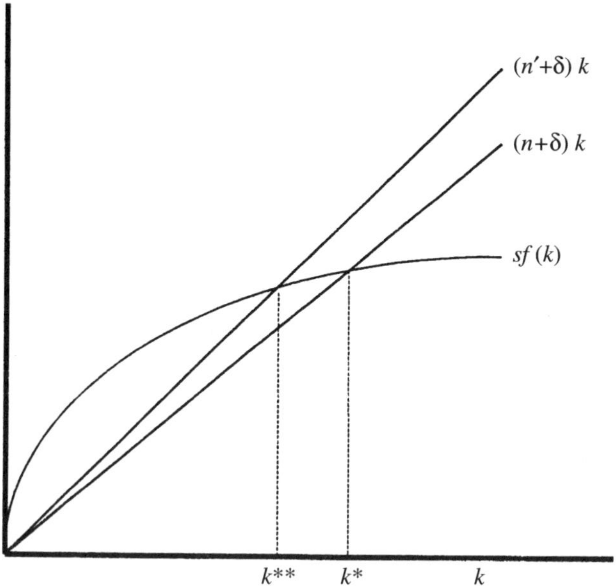

Figure 2.9 Effects from an increase in the population growth rate.

'capital widening' effect produced by an increase in n lowers the long-run value of k to a level that can be sustained by the given propensity to save, s.

Hence the Solow model predicts that countries with higher rates of population growth should tend to have lower levels of per capita income. Although such relation is confirmed by the data, one may question the direction of causation: the causation mechanism, many would argue, could very well run in the opposite direction, from income levels to population growth.

We now introduce technology A in the production function and assume that it grows at a constant exogenous rate. Since in this model technical progress is defined as anything that allows us to produce more output with the same amount of inputs, it is analytically equivalent to an increase in input ('factor augmenting'). The production function is

$$Y = F(K, L A) \tag{2.12}$$

where LA is labour measured in 'efficiency units',[12] and Equation (2.10) can be rewritten as

$$\dot{k} = sF(k, A) - (n + \delta)k \tag{2.13}$$

hence the rate of growth of k is given by

$$\frac{\dot{k}}{k} = \frac{sF(k, A)}{k} - (n + \delta) \tag{2.14}$$

Constant $\dot{k}/k$ in steady state implies that $F(k, A)/k$ must be constant and this in turn implies that A and k grow at the same rate. Therefore, in steady state, per capita capital (and output) must grow at the same rate as A, that is at the exogenous rate of growth of technical progress.

The neoclassical growth model therefore describes an economy where both the capital–labour ratio and labour productivity grow at the constant exogenous rate of growth of technical progress. This result is consistent with (one of) Kaldor's 'stylised facts' of economic growth, namely with the fact that both capital and output per worker grow at an approximately constant rate over long time spans. Changes in technology, the propensity to save and the growth rate of population in this model have an impact on the level of the per capita variables, but not on their growth rate.

In summary, the implications of the Solow model which are relevant to the population and economic development debate can be summarised in four points: (i) an increase in the population growth rate has a negative, 'resource shallowing' effect that depresses the level of per capita income (although we should see no correlation between the long-term growth rates in per capita income and those of population); (ii) since the propensity to save also has a positive impact on the level of per capita income, it becomes important to determine the influence of the population growth rate on the fraction of income saved; (iii) exogenous technical progress explains the growth of the capital/output ratio and of labour productivity, but it does not modify conclusion (i); (iv) this conclusion would instead be modified if there were a link between population growth and technical progress, that is to say by endogenous, population-driven, technical progress.

3 'Revisionism'

Theoretical background

'Revisionism'

In a 1988 survey, Birdsall classifies the different positions within the population–development debate up to the 1980s into pessimists – a category already analysed – optimists and revisionists. The main arguments used by population optimists – of whom the most notable representative is Julian Simon (1977, 1981) – are economies of scale in production and consumption and the belief that population spurs technological change. Both Birdsall (1988) and Kelley (1988a, b) hold that most convincing arguments in support of the idea of population-induced innovations have been offered for the case of agriculture, by the models discussed in Part II of this book. The overall judgement however is that 'the arguments of the optimists, with the possible exception of the advantages of greater population density in rural areas, though of intuitive appeal are as poorly supported empirically (and as intrinsically difficult to support) as are the arguments of the pessimists' (Birdsall 1988: 493).

The 1980s saw the emergence of a position that is referred to as 'revisionism', to stress the fact that most analysts do not subscribe to either population pessimism or population optimism: instances of important 'revisionist' documents are the National Research Council 1986 report and the World Bank 1984 report. This development was linked to a number of factors, many of which are of an empirical nature (Hodgson 1988; Kelley 1988a). The first is the positive demographic and economic trends in the less developed countries during the 1970s. The 1986 National Research Council Report, notices in its first pages that 'despite rapid population growth, developing countries have achieved unprecedented levels of income per capita, literacy and life expectancy over the last 25 years . . . Furthermore . . . there is no statistical association between national rates of population growth and growth rates of income per capita' (National Research Council 1986: 4). In addition, the world's growth rate of population reached a peak in the late 1960s and has declined since, as birth rates declined rapidly in most developing countries (although, as noted in the introduction, such decline does not translate into decreasing 'additions' to population).

The lack of statistical association between population growth and per capita income growth – which had already been pointed out in the 1960s by Kuznets (1967) and Easterlin (1967) – held for the 1970s and 1980s. According to Kelly and Schmidt (1994: 1) 'No empirical finding has been more important to conditioning the "population development" debate than the widely-obtained statistical result showing a general lack of correlation between the growth rates of population and per capita output'.[13]

Research also challenged the theoretical foundations and the empirical relevance of the Coale and Hoover conclusions regarding both the impact of the burden of dependency on savings and its impact on the composition of investments.

However, the emergence of a new approach in the population–development debate was also strongly influenced by a change in theoretical attitude: revisionism is a perspective which highlights market-induced feedbacks to population change in several areas, including technology, the population–environment nexus and institutional evolution. According to Kelley and Schmidt (1994: 9) the analytical foundation of modern revisionism is a perspective which acknowledges the resource shallowing effect of population growth but reckons that population also stimulates the growth of other factors and/or technology (a 'resource-augmenting' effect): labour productivity growth is increased or decreased depending on the relative resource-diluting versus the positive resource-augmenting effects.

Birdsall (1988) stresses the strong use of the logic of the 'invisible hand' in the analysis of the population–development links 'The revisionists view population change as the aggregate outcome of many individual decisions at the micro or family level, and thus one aspect of a larger complex system. The micro- or family-level decisions are made in response to signals provided by the larger system; under the Smithian logic of an invisible hand, these family decisions should be presumed to maximise not only individual welfare, but also social welfare, unless there are clear market failures' (1988: 493).

The invisible hand was also evoked, and not sympathetically, by Paul Demeny, in his presidential address at the 1986 meeting of the Population Association of America: 'The revisionists believe that the invisible hand can work, indeed has been working, its magic not only in the ordinary market place but also in shaping population processes' (1986: 474). Demeny argues that, on the contrary, population processes and most notably fertility behaviour must be an object of attention for governmental action because externalities are attached to demographic behaviour. As for the development of institutions that allow the internalisation of such externalities

> Certain institutional arrangements can diminish the severity of this . . . That is likely to be the case when institutions are the product of a long evolution during which socio-economic conditions changed gradually . . . But short of good luck, the pervasive and multifarious, if typically indirect, influence demographic change exerts on all aspects of the social and economic system virtually guarantees that externalities generated by demographic behaviour are present and significant in any society, particularly under conditions of rapid socio-economic change.
>
> (Demeny 1986: 481)

Finally, both Kelley and Hodgson stress the impact of a changing political and ideological climate.

> The actual declines in fertility occurring during the 1970s could be viewed as finally documenting the plausibility of orthodoxy's solution . . . China's experience proved that government-sponsored fertility control efforts could work even in predominantly agrarian societies . . . With strong institutional support, orthodoxy could have held sway within American demography. The optimistic trends of the 1970s need not have proved more antithetical to its hegemony than the pessimistic ones of the 1950s and 1960s.
>
> (Hodgson 1988: 560)

This was not the case partly as a result of a change in the political (and funding) environment, with a return to traditional views about the family, and challenges to government's family-planning policies, especially those relating to abortion, issues which mobilised social conservatives, often religiously motivated (Kelley 1988a; Hodgson 1988).

In what follows, several strands of analysis that provide theoretical grounds for a 'revisionist' approach to the population–development link are discussed, namely: the contribution of new institutional economics; the analysis of common property and of the 'tragedy of the commons' argument; the population–savings nexus and the population implications of recent neoclassical growth theory. The choice of topics is necessarily selective, and their treatment concise. The purpose of the chapter is to highlight some of the main theoretical issues in the population–development debate, from an economist's perspective, and to provide a framework for the analysis of Part II.

New institutional economics

The debate

Institutions – or, in Williamson's definition, 'organisational variety' – are not accounted for in a standard neoclassical framework where 'firms are production functions; markets are signalling devices; contracting is accomplished through an auctioneer; and disputes are disregarded because of the presumed efficacy of court adjudication . . . The economic purposes served by organisational variety do not arise within – indeed, are effectively beyond the reach of – this framework' (Williamson 1986: 171).

The 1970s saw the emergence of a theory – generally referred to as 'new institutional economics'(NIE) – which sought to extend the neoclassical approach in order to explain institutions other than the market. The discussion of several aspects of the population–development nexus will require reference in particular to one strand of this theory, whose explanation of institutions is based on the notion of transaction costs and property rights. Institutional arrangements, in this view, are deliberately chosen by individuals on the basis of efficiency criteria. Hence, the emergence and evolution of institutions is viewed as the result of rational responses to changes in the underlying economic conditions.

If these claims – whose limitations are discussed below – were sound they would be a source of great optimism, in general, and in particular for the population–development nexus. Population growth indeed induces changes in the structure of relative prices – e.g. the increase in the relative price of land. If societies were always able to produce the right institutional responses there would be less reason to worry about a wide array of issues, including technological adaptation and resource depletion.

Traditional microeconomic theory does not consider the set of activities that normally precede, accompany and follow market transactions, and the associated costs – transaction costs. Within the NIE the use of the concept of transaction costs has become the centre of the analysis of the firm (in Williamson), has enriched the analysis of externalities and is the basis of an approach to the theory of institutions linked mainly to the work of Alchian, Demsetz, North and Thomas. The basic intuitions of such an approach support some of the adjustment mechanisms to population growth described in Part II of this book, and especially the idea that property rights in land evolve in response to demographic growth.

In *The Rise of the Western World*, North and Thomas give the following definition of transaction costs: 'The transfer of goods between economic units requires the provision of information about the opportunities for exchange, or search costs, the negotiation of the terms of the exchange – negotiation costs – and determining procedures for enforcing the contract – enforcement costs. The costs of providing all the services involved are called here transaction costs' (North and Thomas 1973: 93)

In a 1973 article, Alchian and Demsetz define property rights as the socially recognised right to use a resource; they assert that every cost and benefit associated with social interdependencies is a potential externality and that 'internalising' such effects refers to a process, usually a change in property rights, that enables these effects to bear (in greater degree) on all interacting persons. Furthermore, such change is not costless; it does involve 'transaction costs' (in the sense of the cost of a transaction in the rights between the parties): property rights evolve only when the gains of such change become larger than the transaction costs involved, and this in turn must be because of a change in relative prices. Hence institutional arrangements are deliberately chosen by individuals on the basis of efficiency criteria and the basic source of institutional change is essential and persistent changes in relative prices.

Another approach to the theory of institutions – the theory of imperfect information – focuses on the consequences of the abandonment of some of the hypotheses of standard neoclassical theory. Institutions are seen as substitutes for missing markets in conditions of pervasive risks, incomplete markets, information asymmetry and moral hazard. This strand of theory has implications for the analysis of technical change in agriculture – since with transaction costs and incomplete markets the structure of asset ownership becomes an important determinant of the rate and bias of technical change – and for the analysis of the linkages between communal tenure systems and credit markets.

Although in the vast debate on new institutional economics the notion of 'mainstream' actually forms, as David (1994) puts it, an 'inconveniently fluid' point of reference, the criticisms to its central tenet, especially in the case of the property rights school, may be summarised in the general allegation that these theories tend to produce an oversimplified and overoptimistic – functionalist – view of the emergence and nature of institutions. Institutions change or new institutional arrangements emerge simply because the change is called for by (and better suited to) the new underlying economic conditions; furthermore their persistence is close to being considered a proof of their optimality (Bhardan 1989).

Within the NIE field some hold less than straightforward functionalist views. Schotter views institutions, in a game theoretic framework, as 'unplanned and unintended regularities of social behaviour' (Schotter 1986: 118) that emerge unconsciously and gradually from the pursuit of individual interests as agents repeatedly face the same types of social problem: institutions are outcomes of human action that no single individual intended to occur. Also, Douglas North has increasingly emphasised in his contributions that political systems have an inherent tendency to produce inefficient property rights.

The functionalist view of the emergence of institutions is subject to the collective action critic, focusing on the role of interest groups, and also to the more radical Marxist critic. The latter holds the institutions change or do not change depending on considerations of surplus appropriation by a dominant class, and that progress towards a more productive institution may be blocked if it reduces the control of surplus by this class.

A related problem is that of optimality of persistent institutions: functionalism 'often unthinkingly implies the application of the market analogy of competitive equilibrium to the social choice of institutions or the biological analogy of natural selection in the survival of the fittest institution' (Bardhan 1989). Again, this analogy applies to varying degrees to different strands of institutional economists. For the imperfect information school the fact that institutions emerge in response to the absence of certain markets[14] does not imply that they perform optimally: on the contrary Greenwald and Stiglitz (1986) show that market allocations with incomplete markets and imperfect information are, in general, constrained Pareto-inefficient. Schotter, adopting H. Simon's view of behaviour, holds that 'the societal rules of thumb that we have called social institutions are really satisfactory rules that solve recurrent problems once and for all but might not always lead to optimal results' (Schotter 1981: 149).

It is indeed indisputable that dysfunctional institutions may persist for long periods of time – like in one of Akerlof's (1984) woeful economic tales, where the caste system is held in equilibrium because the punishment for breaking the rules is likely to be severe – and that past history shapes institutions. The need to consider the importance of history is vividly expressed by Matthews

> Complexity and inertia reinforce each other. Complex arrangements are difficult to alter radically, so they foster inertia. Inertia makes it easier to respond to changing circumstances and incorporate new institutional ideas, by patching up existing institutions and so making them more complicated still, rather than by starting again from scratch. As a result, each new step in the process of institutional change is determined by its starting point and itself in turn contributes to shaping later developments. History matters. Since the choice at each step is also likely to be affected more or less strongly by stochastic elements, the process assumes some of the character of a random walk. Institutional change acquires a life of its own.
>
> (Matthews 1986: 915)

David (1994) however points out that, although evolutionary change tends to be considered as an obvious attribute of institutions, its explanation requires a radical departure from the 'teleological' explanatory mode, by which the present shape of things can best be explained by considering their function. A powerful insight for understanding path-dependent processes is the idea of self-reinforcing mechanisms. Arthur (1988) describes problems in which multiple equilibria are possible, early history – in part the consequence of small events and chance circumstances – (path dependency) can determine which solution prevails and the actual outcome may be inefficient and difficult to exit from.

The remainder of this section discusses a version of the property rights approach applied to economic history – specifically to the rise of 'modern' Europe – which assigns to population growth the role of the 'predominant parameter shift' inducing the institutional innovations that account for such a rise.

North and Thomas: a neoinstitutionalist view of the impact of population on modern European growth

In *The Rise of the Western World* (1973) North and Thomas use the conceptual framework of neoclassical institutionalism to explain modern European growth. This historical interpretation adopts as its basic reference the neoclassical model, where change occurs via changes in relative prices redirecting factors of production into their most profitable use, through the behaviour of maximising individuals. Within this framework, which in its most abstract form includes no organisations or institutions except for the market, North and Thomas assign a fundamental role to the notions of property rights, transaction costs and contracts.

The starting point of the reasoning is the analysis of the manor economy, or rural seigneurie, which for more than a thousand years was one of the most important institutions of European history. The nature of the institution, and specifically the nature of property rights in man and in land that characterised it, are masterfully described in Marc Bloch's (1971) contribution to the *Cambridge Economic History of Europe*

> The . . . manorial system was not based on slavery in the true sense of the word. Whatever their legal status may have been, even if it went by the name of serfdom, the peasants who composed a seigneurie were in no sense human livestock, fed by their master and owing the whole of their labour to him. They lived on the produce of fields that they cultivated on their own account, which were usually handed down from father to son; and if the opportunity occurred they could sell or exchange the produce in order to procure other necessaries of life. They usually formed little rural communities with a strong *esprit de corps*; exercising common rights over waste land where their flocks could graze and they could gather food; able to regulate the arable land itself in the common interest . . . But they did not work only for themselves, or for the Church and ruler: a great part of their toil went towards the maintenance of one who stood immediately above them.
>
> To this lord, as they called him, the cultivator of the soil owed, first, a more or less important part of their time; days of agricultural labour devoted to the cultivation of the fields, meadows, or vineyards of his demesne; carting and carrying services; and sometimes service as builders or craftsmen. Further, they were obliged to divert to his use a considerable part of their own harvests, sometimes in the form of rents in kind and sometimes by means of taxes in money . . . The very fields that they cultivated were not held to be theirs in full ownership, nor was their community – at least in most cases – the full owner of those lands over which common rights were exercised. Both were said

> to be 'held' of the lord, which means that as landowner he had a superior right over them, recognised by dues owed to him . . . Finally, the lord . . . not only was . . . a rentier of the soil and a beneficiary of the services; he was also a judge, often – if he did his duty – a protector, and always a chief whom . . . those who held their land from him or lived on his land were bound, by a very general but very real obligation, to help and obey.
>
> (Bloch 1971: 235–236)

This system prevailed all over medieval Europe in conditions of land abundance, labour scarcity, and weak central authority. Within the territory of the manor only a small portion of the land was actually cultivated, depending on the availability of the scarce factors – labour and capital – and the crops would be typically rotated according to a two fields system.[15] Peasants cultivated fields for their own account and used the pasture and wastelands essentially in return for labour on the land of the lord and for other dues of various kinds. Land under the lord of a manor would be granted under different types of tenancy, with the main distinction being the one between tenants who could at any time – at least in principle – return possession to the lord and depart (free tenants) and tenants who were bound by law to remain on the land, this latter group constituting the majority of the population.

North and Thomas interpret the manor economy as being essentially based on a contractual arrangement whereby labour services were exchanged in return for the lord's protection. 'The fortified castle and armoured knights . . . having specialised skills in warfare, provided local security which could never be equalled by any group of peasants ill-armed with primitive weapons and lacking military skills' (North and Thomas 1973: 19), while the lord and his knights, who specialised in protection and justice, depended on serfs for what they consumed. Hence, serfdom in Western Europe, they maintain, was essentially a contractual arrangement, that is an agreement – between parties involved in governing a transaction – which cannot be unilaterally changed by either party (North and Thomas 1971).

The specific contractual arrangement chosen was the one that minimised costs (transaction costs), given the overall economic environment. In the absence of an organised product market, as in the high Middle Ages, all modern forms of contracts in agriculture – fixed rent, fixed wage and sharecropping – implied very high transaction costs: for example the absence of an organised market for goods, providing information on prices, would make it extremely difficult to establish the value of the mix of goods to be eventually provided in an output sharing agreement. The input sharing agreement of the manor economy involved higher enforcement costs but much lower transaction costs and was therefore a 'fully rational' contractual arrangement for such an economy.

The gradual breakdown of this system was, according to *The Rise of the Western World*, essentially a consequence of the impact of population growth on relative prices and institutions. The theoretical structure of the argument is based on the notion (Alchian and Demsetz 1973) that property rights evolve – 'internalising' the externalities – only when the gains of establishing them become larger than the transaction costs involved.

Given the assumption that individuals are maximisers, North and Thomas maintain that economic growth will occur only if property rights provide private incentives to undertake the socially productive activities and, therefore:

> No sustained economic growth could be set in motion until fundamental institutional developments created or simulated or approximated private property on land and a free labour market.
>
> (North and Thomas 1971: 778)

> The creating, specifying and enacting of such property rights are costly, in a degree affected by the state of technology and organisation. As the potential grows for private gains to exceed transaction costs, efforts will be made to establish such property rights.
>
> (North and Thomas 1973: 8)

Hence, population growth is viewed as the explanation for growth because it was 'the predominant parameter shift which induced the institutional innovations that account for the rise of the Western World' (ibid.) by making their potential benefits larger than their costs.

As the local manor became overcrowded, new land was settled: population growth produced a frontier movement. The settlement of lands in Northwest Europe, with different soils and climate, led to the emergence of varied patterns of agriculture. Such differentiation increased the profitability of trading with two major consequences: towns revived or developed as natural centres for trade, and as centres for the production of manufactured goods; the increasing gains from trade established the incentive to extend the protection of commerce – property rights – beyond the confines of the single manor.

The influence of the population variable on the labour market is more ambiguous. On the one hand, in a neoclassical institutionalist model, individuals or groups negotiate, or renegotiate contracts on the basis of their relative strength, which in turn is linked to relative scarcity – and the resulting institutional dynamics does not differ from the one postulated by the demographic model. Hence *The Rise of the Western World* holds that the master–servant aspect of manorialism gradually fell away after the Malthusian checks of the fourteenth century, when the dramatic population decline and the consequent scarcity of labour improved the bargaining strength of the workers. On the other hand, however, North and Thomas also assert that it was population growth, through its impact on the establishment of organised markets for goods, that indirectly caused the dismissal of the input-sharing arrangement which characterised the manorial economy.

The thesis is that the establishment of organised markets for goods, the concomitant use of money as the account unit and the emergence of market prices, lowered the transaction costs of establishing a system of wages, rents or shares by contract, which was to replace the contractual arrangement of the classic manor (i.e. the provision of labour services in return for protection) since, *ceteris paribus*, free labour is more productive than that of serfs which is not voluntary. Hence 'Although seemingly fallen into disrepute, the traditional explanation for the decline of the manor – namely the rise of a market economy – when considered in the context of the preceding analysis is sufficient to explain the eventual disappearance of the essential element of the manorial system: the sharing of inputs in the form of labour dues' (North and Thomas 1971: 794).

The transition from the complex tenant–lord relation of the manor to the modern fee-simple absolute ownership of land is seen as the product of the same economic forces that created a free market for labour.

> Feudal law did not recognise ownership in land. Rather, its basic characteristic was that several persons could have jurisdiction over a given piece of land. King, lord, and peasant (in somewhat different senses) each held a right to the same piece of land . . . Common law grew up around real actions which brought litigation before the King's court. This court had jurisdiction over all freeholders, in contrast with the Court of the Lord – the seigneurial court – which had jurisdiction merely over unfree tenants . . . The various forms of action which evolved over time gradually gave protection to the freeholder in the possession of

> land and its hereditary transfer . . . the alienability of land took somewhat longer . . .
>
> . . . The long history of the gradual transformation of villein to copyholder, through intervention by the chancellor and ultimately to protection of common law, followed a similar pattern of gradual evolutionary changes in the rights of villeins to heritability and alienability . . . The Action of Ejectment in Elizabethan times was the final stage in common law remedies which embedded the tenants' rights in the fundamental institutional structure.
>
> (North and Thomas 1971: 801–802)

Thus the pattern of change of English common law is seen as a rational response to economic pressures, and particularly to the scarcity of labour in the fourteenth and fifteenth centuries.

> In Western Europe the most effective way to retain tenants was to lower rents and to relax servile obligations [which] led to the innovation of lengthy leases, which soon came to be life leases . . . Life leases turned out to be a last ditch effort by lords to retain their customary rights by signing them away for only one generation of tenants. But since recurrent plague did not allow the population to expand for several generations, these agreements themselves took on the force of custom and eventually the tenants obtained by customary practice the right of inheritance. In the late fifteenth and sixteenth centuries such arrangements came to be considered as equal in law to copyholds and as subject only to the now-customary encumbrance of a fixed money payment or quit rent. A secular rising price level during the sixteenth century reduced this to a purely nominal payment by the year 1600. The manorial economy thus met its death: labour services were now irrevocably replaced by money rent payments; land was now tilled by free tenants and/or by workers receiving money wages, who were free to seek their best employment.
>
> (North and Thomas 1971: 798–799)

The final result of this institutional evolution was fee-simple absolute ownership of land and a free market for labour, which are taken to be two essential preconditions for efficient resource allocation, and hence for economic growth.

In summary, the development and expansion of a market economy is interpreted as a direct response to the opportunity to gain from the specialisation and trade made feasible by population growth. Furthermore, the change in the labour and land markets are seen as a consequence of the development of the markets for goods and of the change in bargaining strength of the labour force after the Malthusian checks of the fourteenth and fifteenth centuries.

The impact of population is on the whole not unambiguous. Population growth is seen as inducing diversification of the economy and the line of reasoning represented in Figure 3.1 leads to the conclusion that population growth contributed crucially to the decline of the input-sharing agreement of the manor and established a firm positive link between population and modern economic growth.

However, in *The Rise of the Western World*, North and Thomas also adopt the interpretation that the tendency to free the peasants and to create more stable rights on the land was indeed associated with the consequences of population decline. Hence the assertion that population growth was the predominant parameter shift that explains modern European growth is justified only if the impact of the development of the market is given much more weight than the impact of the change in bargaining strength of labour.

It is interesting to note that North and Thomas give, *passim,* an interpretation of the transmission of a major technological innovation in agriculture – the three fields system[16] – which is very close to the idea of endogenous population-induced technical progress in agriculture in a model by E. Boserup, discussed in Part II. Although the system had been known since the late eighth century, and although it

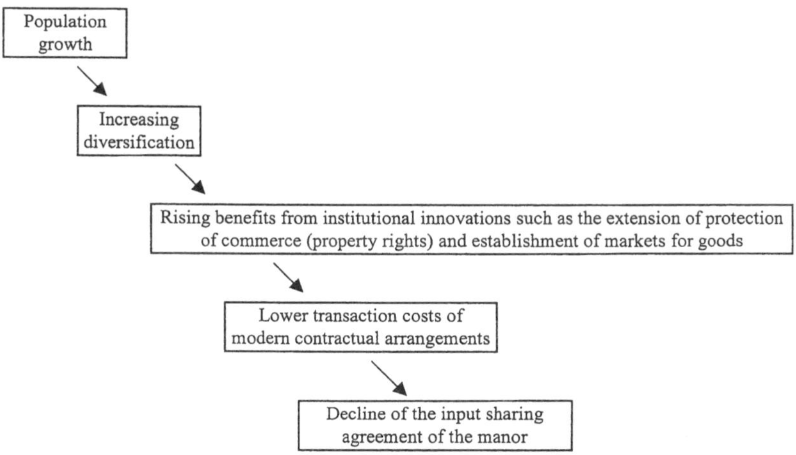

Figure 3.1 Population and growth in *The Rise of the Western World.*

had several enormous advantages – including the fact that the arable land bearing crops was increased substantially, labour requirements were spread more evenly through the year and having two harvest periods reduced the chance of famine due to seasonal crop failure – its spread throughout Western Europe took several centuries. Such slow adoption would be explained by the fact that the system became a more efficient form of organisation only when a growing population led to diminishing returns to labour and land had to be used more intensively.

The Rise of the Western World interprets the history of the whole of Europe between 900 and 1500 along the lines of a single model. It is, according to the authors, from the 1500–1700 period that economic events must be examined according to national boundaries, as in this period institutions and property rights within the emerging nation-states took divergent paths. The divergence is revealed by the analysis of population trends. During the sixteenth century all countries in Western Europe experienced similar trends of growing population and diminishing returns. A reversal occurred in the seventeenth century, but this time the Malthusian checks, unlike those of the fourteenth century, hit different countries in Europe with differing intensity. Population declined in the Spanish Netherlands, Spain and perhaps Portugal and Germany, stagnated in Italy and France, but actually increased in the Dutch United Provinces, England and Wales.

North and Thomas surmise that it was the different efficiency of economic organisation among countries that played a large part in determining the effectiveness of the Malthusian checks; such difference would be explained by the type of property rights created by the emerging states in response to their continuing fiscal crises.

> Governments take over the protection and enforcement of property rights because they can do so at a lower cost than private volunteer groups. However, the fiscal needs of government may induce the protection of certain property rights which hinder, rather than promote, growth; therefore we have no guarantee that productive institutional arrangements will emerge.
>
> (North and Thomas 1973: 8)

One example of inefficient property rights discussed at length is the fiscal policy of the French Crown, which barred the spread of an extensive market by establishing a system of internal tariffs, and perpetuated and enforced guild monopolies in return for revenues. The lack of a major enclosure movement in France would be the consequence of the lack of incentives, because of a limited market, and of the costs of changing a legal system clogged with bureaucrats.

Another example of inefficient property rights is the perpetuation of the Spanish Mesta – the right granted to the shepherds' guild to drive their sheep across the country. As population grew and land became scarce, the social rate of return of improving the efficiency of agriculture rose, but the private one did not, since landowners were constantly exposed to the risk that their crops could be eaten or ruined by flocks of migrating sheep.

> The King of Spain derived a substantial part of his revenues from the Mesta, and it was not clear that he could gain from abrogating their rights. Although the income of society would have been increased by such a change, it would appear that the Crown's own revenue from land taxes, reduced by the costs of reorganising property rights and collecting the levies, would not, at least in the short run, have equalled the traditional revenues from the Mesta.
>
> (North and Thomas 1973: 4)

Hence the maximisation of the present value of state income frequently led to the formation of property rights that actually damaged economic growth. In a 1989 article North stresses:

> One of the most evident lessons from history is that political systems have an inherent tendency to produce inefficient property rights which result in stagnation or decline. There are two basic reasons for this result. First the revenue that can be raised by rulers may be greater with an inefficient structure of property rights that can, however, be effectively monitored, and therefore taxed, than with an efficient structure of property rights with high monitoring and collection costs. Second, rulers can seldom afford efficient property rights. That is, even when rulers wish to promulgate rules on the basis of their efficiency consequences, survival will dictate a different course of action, because efficient rules can offend powerful interests in the polity.
>
> (North 1989: 1321)

The interpretation of modern European growth offered by North and Thomas in *The Rise of the Western World* provides many powerful insights; still it is very controversial and suffers some of the basic limitations of neoclassical neoinstitutionalism.

Historians have expressed doubts about what has been seen as the superimposition of an 'exogenous' theory to historical facts (Wilson 1977). The view according to which the manor economy was based on a contract between individuals trading different 'factors' neglects the power structure of society and begs some fundamental questions. For the majority of peasants, certainly for those unfree peasants who inherited their status, the tenancy–lord relationship lacked one essential characteristic of a contract, the fact of being a voluntary agreement. As for the other peasants, their condition is described namely by Marc Bloch in these terms: 'the "free" man is the man who can choose his own lord – as a vassal does, whose homage must be renewed as lord succeeds lord, under pain of losing his fief no doubt, but in theory of his own free will' (Bloch 1971: 254).

The tenant–lord relationship was the result of a process by which individuals using some comparative advantage (in violence or in military skills) and/or the occurrence of a gift of land from someone (e.g. the king) became increasingly capable of obtaining labour-cum-land, from the men or communities which submitted themselves to him, in exchange for protection. He could at the same time protect them efficiently, and force them into the relationship. Brenner argues that North and Thomas' view of an exchange between individuals with different endowments 'only begs the fundamental question of class: how do we explain, in the first place, the distribution of the land, of the instruments of force, and of military skill within societies' (Brenner 1976: 35).

The tenant–lord relationship is actually described by Bloch as the result of a long historical process encompassing a great deal of violence and manipulation of the law to the lords' advantage.

> Since the state got weaker and weaker . . . the lords kept those judicial powers that had been given them and usurped all or part of the rest . . . Now in this way the seigneury acquired a powerful instrument of consolidation and expansion, – not merely through the bare right of judicial decision, but also, and perhaps mainly through the confusion of this right with the right to issue orders and punish those who disobeyed; in Frank terminology, the ban. This valuable right had originally been reserved to the king and his representative . . . counts, or their subordinates, . . . went so far as to force the unhappy and almost defenceless freemen to work like corvèable dependants in their fields and vineyard and meadows. Many a group of men was annexed to a seigneurie in this lawless fashion . . . But the working of the immunities had

> far wider and far more durable results than this. Among those who lived on immune land, or those who though living outside it had commended themselves to its lord, a great many had at the outset been very loosely bound to the lord and owed but little to him. The ban allowed him to stiffen up both the relationship and its practical burdens.
>
> (Bloch 1971: 262)

And, in another part, the question of the voluntary nature of contract is explicitly touched:

> The medieval contracts of subjection regularly purport to be inspired by the free will of the new subject and especially, when the lord is a church, by piety. But in social life is there any more elusive notion than the free will of a small man? . . . There were many other forces at work to make the small man pliable; from hunger – sometimes a declared cause, but generally in the case of the landless worker – to the wish to share in those common rights which a lord reserved for his dependants; up to that sheer oppression, about which the written contracts are of course chastely silent, but which many other sources disclose. . . . In this troubled society, whose central authority could not get into effective touch with the masses, violence helped to transform social conditions the more effectively because, through the play of custom, an abuse might always by mutation become a precedent, a precedent a right. . . . As only a master already strong could protect a man effectually; as only a prominent personage of this kind could put decisive pressure on a man (we must always consider heads and tails) – the protector of lands or body was generally an individual, or religious institution, already protecting other dependants in the same fashion. So a seigneurie, once only modest nucleus, threw out long tentacles on every side.
>
> (Bloch 1971: 268–270)

Finally, certainly the tenants paid rents, and received the public good of defence, but how were those prices established? And what was the relationship between the distribution of assets underlying the contract and the lack of progress in medieval agriculture? As Genicot observes for the Malthusian interpretation, the assumption of declining productivity in agriculture – which is crucial for the model – is a reasonable one for most pre-industrial European economies. There is ample historical evidence regarding the thirteenth century of expansion of cultivation to less fertile soils and loss of fertility of other soils, because of lack of fallow periods or manure and because of soil erosion and desiccation, due to the removal of the forest cover. The alternative to Malthusian diminishing returns – i.e. intensification based on the introduction of better techniques – could not materialise in a system where there was hardly any investment in agriculture, where peasants were poor because assets were unequally distributed, charges heavy and often increasing with the ability of the tenant to pay, while the surplus which accrued to the clergy and nobility was deflected into unproductive expenditure (Genicot 1971: 670–671).

Brenner has sharply criticised the interpretations of long-term economic change in late medieval and early modern Europe which rest solely on 'objective' economic forces, in particular demographic fluctuations (the Malthusian model with its cyclical dynamics) and the growth of trade and markets (the commercialisation model). He considers the North and Thomas approach as a radical formulation of a type of perspective which tries to explain long-term economic development in terms of changing institutionalised relationships of 'equal exchange' between individuals trading different 'factors' under changing market conditions (Brenner 1976: 31).

In the 1976 essay which opened the 'Brenner debate'[17] Brenner contends that these attempts at economic model building fail to explain historical developments: similar demographic trends throughout Europe between the twelfth and the eighteenth centuries yielded very different results in terms of development and, similarly, the force of the market was associated both with a decline of serfdom and with its reinforcement.

In Genicot's (1971) account, the first reaction of most of the lords to population decline and to the problem of labour scarcity was to try to impose a solution by force.

> On the morrow of the plagues, the pressing need was to cultivate the abandoned holdings. To meet this need the landlords went back, or sought to go back, on the freedom to abandon holdings granted or tolerated during the period of demographic and economic expansion . . . made arrangements with each other to facilitate the search for and restitution of fugitives; sought to extort the maximum work for the least wage from their tenants . . . demanded of them the effective performance of every corvée, even those which had previously converted into money payments

> . . . From Spain to Norway, princes, parliaments and city magistrates vied with each other to regulate wages.
>
> (Genicot 1971: 705–706)

But after this first reaction, which was nearly universal in medieval Europe, it appears that a single model fails to explain the paths that the tenant–lord relationship took in Eastern and Western Europe, for in the former the land-owning class succeeded in reinforcing serfdom while in the West peasant opposition and landowners competition for labour ultimately led to a deep transformation of that relationship to the tenant's advantage.

The logic of the demographic model and of neoclassical institutionalism would dictate that following 'the demographic catastrophe'; in particular that of the late fourteenth and fifteenth centuries, peasants used their economic positions (their scarcity) to win their freedom. Yet the decline in population has also been used to explain the reinforcement of serfdom in Eastern Europe by landlords who were protecting their own source of income.

In neoclassical institutionalism, change in relative prices leads to the renegotiations of contracts, and the direction of such renegotiations is predictable on the basis of the economic forces that initiate it. Brenner denies such predictability and contends that the different historical outcomes are explained by the historically specific patterns of development of the contending agrarian classes, and their relative strength, in different European societies. 'It is the structure of class relations, of class power, which will determine the manner and degree to which particular demographic and commercial changes will affect long-run trends in the distribution of income and economic growth, and not vice versa' (Brenner 1976: 31).

For example, the extraction of rent depended on the character of the lord–peasant class relation: free tenants would pay a fixed, customary rent; serfs could be forced to pay additional, arbitrary rents. Thus, although economic forces, such as those arising from demographic change, would tend to increase the landlords' power to extract rent, the extent of such increase depended on the nature of class relations, while these latter are relatively autonomous from economic forces. Brenner maintains that the bargaining power of a group is not just a function of relative price change, but also of other, sociological and political factors: the relative strength of the agrarian classes in different European societies is viewed as a function of '. . . their relative levels of internal solidarity, their self-consciousness and organisation, and their general political resources – especially their relationship to non-agricultural classes (in particular, potential urban classes allies) and to the state (in particular, whether or not the state developed as a "class competitor" of the lords for the peasant surplus)' (Brenner 1976: 52).

Brenner's view is shared by the authors of an extensive analysis of agrarian relations across developing areas. Binswanger *et al.* (1995) observe that in most countries at some point in time and for long periods in history a class of overlords has used violence or the threat of violence and control over the state to keep farmers within their estates – through coercion or, at a subsequent stage, by depressing the expected utility or profits of the free peasants.

Population and the environment: local commons

What is discussed under this rubric is solely the issue of local commons, since these represent an important class of environmental problems in the developing countries and since arable land problems belong to this class. A further delimitation of the discussion is that it tracks the structure of the 'tragedy of the commons' argument, which is closely associated with population pessimism. This argument begins with a view of communal property[18] as open access. If the interaction among users of such property is then represented as a prisoner's dilemma, when population grows resource depletion is unavoidable. Hence the issues taken up here are two: the nature of communal type ownership and the possibility of cooperation in common property resources (CPR) management.

The conventional analysis of communal property and the ensuing policy prescriptions have been largely based on the views of the property rights school. Alchian and Demsetz (Demsetz 1967; Alchian and Demsetz 1973) draw a sharp distinction between private property and communal property, the latter being described as an open-access situation and therefore associated with great externalities. Communal ownership is defined as a situation where the use of the resource is determined on a first-come, first-served basis, and communal property on land is actually equated to 'the use of a city sidewalk' (Alchian and Demsetz 1973). In his 1967 essay, Demsetz admits that in principle members of a community could establish and enforce rules that limit individual use of resources, but considers this an unlikely outcome, since negotiating and policing costs would be too high. Indeed, in a world of maximising individuals, the probability of free riding increases in proportion with the number of individuals involved, and so do enforcement costs.

The first argument against a communal type of rights is thus a resource degradation argument: when resources are under group control, in the absence of collective action in the form of rules limiting individual use, aggregate consumption of the resource will tend to be inefficiently high, resulting in degradation – an outcome that, following Hardin (1968), is often referred to as the 'tragedy of the commons' – and investment in the maintenance of the resource inefficiently low. A classic example is that of a common grazing field of fixed size. Each individual has two strategies: grazing at a level that yields a high payoff but imposes a cost to others because of overuse of the resource ('defect'), or 'cooperate', limiting individual use in such a way that the resource is preserved for all. If the structure of payoffs is such that each player is better off when defecting, independently of what the others do, all will defect (Runge 1992).

This argument has been criticised both on theoretical and on empirical grounds: ample economic literature has explored the possibility of cooperative behaviour, and especially its emergence when self-interested individuals are involved in repeated games; furthermore empirical evidence from developing-country experience offers many examples of balanced management of common property resources within small local communities.

Both empirical and theoretical arguments suggest that cooperation is hard to sustain and that without effective institutions common pool resources will be under-provided and overused. At the same time there is considerable empirical evidence showing that in many cases relatively informal collective management of common property resources – involving the constitution and enforcement of rules by appropriators of the resource – can avoid the 'tragedy of the commons'.

In a game theoretic framework, the 'tragedy of the commons' problem has often been represented as a game with a 'prisoner's dilemma' structure. Suppose the CPR game is represented as a game with two players x_1 and x_2 each with two strategies – to invest a productive input (labour, or capital) in the CPR ($x_i = 1$) or to invest it outside the CPR ($x_i = 0$). Output from the CPR is a function of total investments by all players, and the payoffs for investing are $F(1)$ if only one of the two players invests in the CPR, $F(2)/2$ if both invest in the CPR. The payoff for not investing is the factor price in the rest of the economy w. The resulting payoff matrix is depicted in Figure 3.2.

		Player 2	
		$x_2=1$	$x_2=0$
Player 1	$x_1=1$	$F(2)/2$, $F(2)/2$	$F(1)$, w
	$x_1=0$	w, $F(1)$	w, w

Figure 3.2 A common property resources game.

Source: adapted from Ostrom *et al.* (1994)

If $F(1)>w$ and $F(2)/2>w$ there is a unique Nash equilibrium, with both players investing in the CPR. This resembles the prisoner's dilemma, since each player has a dominant strategy. The outcome is suboptimal and eventually 'tragic' if F is strictly concave[19] (Ostrom *et al.* 1994).

The nature of the game is non-cooperative – the players are not permitted to make binding agreements and equilibrium obtains when each of them independently chooses a strategy in a way that maximises his payoff. Each player acts independently but takes the probable choices of other players into account in choosing a strategy. Any pair of strategies in which each player maximises his payoff given what the other player does is called a Nash equilibrium. In the prisoner's dilemma the non-cooperative solution is the dominant strategy for each player, i.e. one player is always better off choosing to defect no matter what the other player chooses, and the resulting equilibrium is not a Pareto-optimal outcome. In general, since Nash equilibrium requires players to choose their best actions, given the action of others, in many applications of one shot games collusive activities – i.e. obeying behavioural norms which are not in the agent's immediate interest – cannot be supported as a Nash equilibrium.

There are however other possibilities. The CPR problem may not have the form of a prisoner's dilemma game: for instance the common management of irrigation systems has been described with reference to games where defecting is not the dominant strategy and cooperation is more likely to arise. Furthermore, cooperation may arise because individual choice is based on some degree of fairness or solidarity with others and/or simply on the perception that social norms work better than individual calculations (Bardhan 1993b).

Economists have mostly tried to explain cooperative behaviour and adherence to social norms, while maintaining the usual hypothesis of 'selfish' utility maximisation in the context of repeated (rather than one-shot) games, since social problems of cooperation generally require repeated interaction. The basic idea is that agents do not deviate from the norm because of the fear of subsequent punishment by other players. Hence, while sociologists and anthropologists look at social norms and codes of conduct in understanding cooperative behaviour, the game theoretic explanations offered by economists generally maintain the usual emphasis on incentives and penalties.

When the same (ordinary) game is played repeatedly by a fixed set of players, the set of equilibria in them is greatly enlarged as compared with the single play of a component game and may include realisations on the payoff possibility frontier that are, in general, unobtainable as equilibrium outcomes in the single play that is being repeated (Friedman 1990; Sabourian 1990).

In the prisoner's dilemma the players can achieve an outcome normally associated with cooperation if they adopt a parallel strategy (trigger strategy)[20] of refusing to confess at first iteration; continue to refuse in each later period if both have in the past also always refused, but revert forever to the policy of confessing if the other fails even once to cooperate.

Hence, with the introduction of time, many types of activities, including cooperation, which at first would seem irrational in terms of 'selfish' maximising behaviour, can be accounted for as solutions to some repeated games.

In general, the sustainability of cooperative equilibria among self-interested individuals depends on a number of conditions, which mostly determine the extent to which the future actually matters. One pertains to the time horizon of interaction: the game must be infinitely repeated, or there should be sufficient uncertainty about how many times it will be repeated.[21] Furthermore, players should not discount future payoffs too heavily and 'other' players must have at their disposal retaliatory strategies that 'hurt' the deviator sufficiently in future periods – even when future payoffs are discounted – and that are credible. Finally, players must also be able to observe with sufficient reliability whether agreements are being kept (Friedman 1990; Sabourian 1990; Seabright 1993).

Although, as Bardhan (1993a) points out, the game theoretic literature cannot account for the impact of many real life, dynamic features of common resources use – e.g. the updating and contingent modifications of the rules of the game, changes in the payoff matrix, the possible evolution of strategy sets as an outcome of successive rounds of a game and the flexibility of the system of sanctions (in contrast with the strict trigger strategies of punishment in the theoretical models) – they offer an economic explanation to the empirical evidence on cooperation in resource management.

Many of the theoretical conditions for cooperation, and considerable empirical evidence, actually refer to local commons i.e. resources owned by small communities, such as villages, and the typical assets considered in the literature are grazing land, inshore fisheries, irrigation systems, and forests. In such contexts both monitoring and the eventual implementation of social sanction are easier, and the likelihood of cooperation is enhanced by the similarity of needs and shared norms. Finally, arable land, which is the crucial resource for agriculture, not only is a 'local common' but also, as will be discussed in Chapter 5, its use is individual – a condition which does not apply to other local commons – and in many cases individuals do receive inheritable rights (Bardhan 1993a, b; Bromley and Cernea 1989; Dasgupta 1993; Seabright 1993).

Population and savings

The Coale and Hoover hypothesis that high youth dependency rates in developing countries reduced savings and therefore represented a hindrance for growth was widely recognised as valid for a substantial period of time and was also confirmed by an extensive cross country analysis by Leff (1969). By the beginning of the 1980s however accumulated empirical research showed little support to this notion, or at least indicated that it was not possible to make a simple general statement about the relationship between population growth and aggregate savings (Kelley 1988b).

The more recent literature has analysed youth dependency effects on savings within the life-cycle model of consumption and savings developed by Modigliani. Results tend to show that there are negative dependency effects on saving at the household level (Kelley and Schmidt 1996).

In the model, because of the hypothesis of convex preferences in intertemporal consumption, consumers are supposed to use their lifetime resources so as to maintain a fairly even standard of living. On the other hand income rises steadily over most of the working life, with a peak between the age of fifty and sixty, and drops after retirement. The resulting lifetime pattern of savings for a 'typical' household is that saving is very low or negative at the beginning of the working life, grows between 40 and 65 and becomes negative after retirement.

Although in this representation a typical household does not accumulate wealth, there will be positive aggregate savings in the economy as long as population and/or per capita income are growing. This is because a growing population implies that young, saving households outnumber old, dissaving households; and because per capita income growth implies that young households have higher lifetime earnings than earlier generations. Both effects are captured by total income growth. Hence there is a positive effect of the rate of growth of income on savings, termed the 'rate of growth effect'.

The life-cycle theory predicts that population and per capita income growth will exert the same 'rate of growth effect'. Although early empirical studies did not find a significant impact of the population growth rate on savings, Mason (1987) argued that this occurred because they failed to take into account the effect of population growth on the youth dependency rate. After controlling for dependency, both terms – population growth and per capita income growth – have an impact on savings.

Specifically, the magnitude of the rate of growth effect would depend on the time lag between the average age of earnings μ_y and the average age of consumption μ_c. These are obtained by weighting each age with the associated values of (respectively) earnings and consumption. Since earnings are concentrated in the young age and consumption is evenly spread, it will be $\mu_y < \mu_c$. If one assumes that consumption during the child-rearing years increases with the number of children, a higher youth dependency ratio leads to a decline of this lag.

Hence the analysis of the effect of dependency on savings can be carried within a model where the saving ratio (*S/Y*) is a function of the rate of growth of income (g_Y, which is the sum of the rate of growth of population and of per capita income) and the interaction between g_Y and the youth dependency ratio *D* (a consumption timing term). The expected signs are positive for the g_Y term and negative for the consumption timing term D^*g_Y (Kelley and Schmidt 1996).

Empirical estimates within such a framework confirm that the youth dependency ratio has a substantial impact on saving and that there is a positive rate of growth effect on savings (Mason 1987; Kelley and Schmidt 1996).

The relationship between population growth and savings is however complex, and difficult to analyse. On the one hand a rise in population growth increases the youth dependency ratio, on the other hand the growth rate of population has a positive impact through the rate of growth effect.[22]

Furthermore, the relationship between youth dependency and consumption is more complex than a simple model such as the one described above posits. Reasons why it is difficult to make simple generalisations about the impact of children on the household include the following: children may substitute for other forms of consumption or child rearing costs may be met by reducing other members' consumption levels;[23] children may influence income levels (family members may increase their work effort while female participation to the labour force is reduced) and may contribute to household market and non market income (Kelley 1988b; Mason 1988).

In summary, recent research about the relationship between population growth and savings in the developing countries suggests less clear-cut conclusions than those reached with the Coale and Hoover approach.

Population in recent neoclassical growth theory

The revival of interest for growth since the 1980s is essentially linked to two disappointments with the Solow growth model. The first is the essential shortcoming that the long-run per capita growth rate is determined entirely by an element – the rate of technological progress – which is exogenous to the model. The fact that the growth rate of population is also exogenous – although in the Solow model the growth rate of the labour force (and of population) is the long-run growth rate of output – is another obvious flaw. However, understandably in their perspective, growth theorists have not paid to this topic the attention devoted to the issue of technical progress. Second, the empirical evidence does not support the model's prediction that countries with lower starting values of the capital–labour ratio have higher per capita growth rates and tend thereby to catch up ('converge') with those with higher capital–labour ratios.[24]

Limiting our brief account to the issue of the determinants of growth, it is easy to show that endogenous growth can be obtained by eliminating the hypothesis of diminishing returns to capital, i.e. by specifying the production function of the Solow model as

$$Y = AK \tag{3.1}$$

where A is a positive constant reflecting the level of technology. Since Equation (3.1) implies that

$$\frac{f(k)}{k} = A \tag{3.2}$$

the growth rate of the capital–labour ratio of Equation (2.11) can be expressed as

$$\frac{\dot{k}}{k} = sA - (n + \delta) \tag{3.3}$$

The relationship between k and its growth rate is represented graphically in Figure 3.3, where the downward sloping $sf(k)$ function is replaced by the horizontal line sA: all per capita variables in the model grow at the same rate, given by the difference between sA and $(n+\delta)$ (Barro and Sala-i-Martin 1995).

The tendency for diminishing returns could be eliminated – while maintaining the assumption of perfect competition – if there were knowledge creation as a side effect of investments. Arrow had made the accumulation of knowledge endogenous already in a 1962 article, by assuming that the growth of A was proportional to the accumulation of physical capital K as a result of learning-by-doing, i.e. by interpreting the accumulation of knowledge as a by-product of mechanisation.

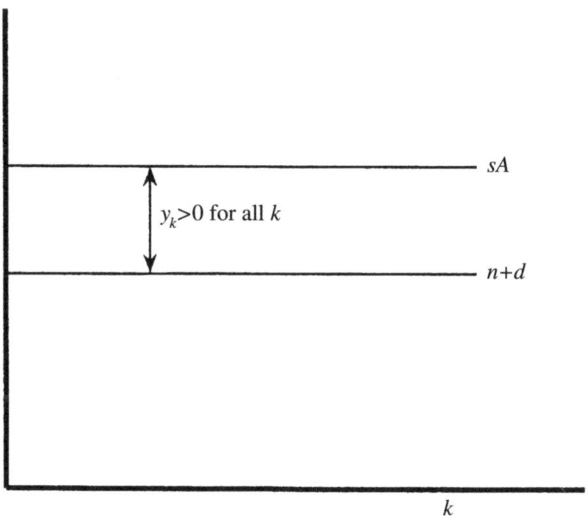

Figure 3.3 The *AK* model.

Source: Barro and Sala-i-Martin (1995)

Romer used a similar framework in a 1986 seminal paper where the condition of decreasing marginal productivity of capital is maintained at the level of the firm, and a positive technological externality, linked to the total amount of capital is introduced at the aggregate level (Romer 1986).[25] The production function for firm *i* is

$$Y_i = F(K_i, A_i L_i) \tag{3.4}$$

The additional assumptions are that an increase in a firm's capital stock leads to a parallel increase in its stock of knowledge A_i; and that once discovered knowledge spills over instantly and all firms can use it at zero cost (i.e. a firm's knowledge is a public good that any other firm can access at zero cost). Therefore

$$A_i = K \text{ for all } i \tag{3.5}$$

Hence the production function for firm *i* is

$$Y_i = F(K_i, KL_i) \tag{3.6}$$

Equation (3.6) is homogeneous of degree one in K and K_i for given L – there are constant returns to capital at the social level – although each individual firm sees the level of technology as given and regards its production function as neoclassical, with constant returns to scale and positive and diminishing marginal products of capital and labour.

It can be shown (Barro and Sala-i-Martin 1995: 147–150) that in this model the steady state per capita growth rate is increasing in the size of the labour force L. Intuitively this happens because the 'engine of growth' At is proportional to the stock of capital K and this, given the capital–labour ratio across the economy, is proportional to L.

Hence the model predicts that, other things equal, countries with a higher level of population will tend to grow faster in per capita terms. However this 'scale effect' of population does not have a clear empirical support. Barro and Sala-i-Martin (1995) suggest that the scale variable for spillovers L does not relate necessarily to aggregates measured at country level: it could, for example, be larger than the size of the domestic economy if producers benefit from knowledge accumulated in other countries.

However this makes the model very difficult to falsify with aggregate macroeconomic data. 'If we are to assume a constant population, then it is the same to model "learning by doing" in terms of increases in the capital stock or in terms of the more mundane idea of mechanisation (that is a rising capital-labour ratio)' (Valdés 1999: 121–122). In this version of the model there are no scale effects and the rate of growth of population enters with negative sign in the determination of the equilibrium per capita growth rate of the economy (Valdés 1999).[26]

The above discussion illustrates the observation by Kremer that

> Models of endogenous technological change . . . typically imply that high population spurs technological change. This implication flows naturally from the nonrivalry of technology . . . [since] the cost of inventing a new technology is independent of the number of people who use it . . . However, despite its ubiquity in the theoretical literature on growth, this implication is typically dismissed as empirically undesirable.
>
> (Kremer 1993: 681)

As a way of further illustration of this point we may refer to a simplified version (as formulated by D. Romer 1996) of models where – like in the Solow and the *AK* model – knowledge accumulation is the driving force of growth, but knowledge production is explicitly modelled. The model is described by Equations (3.7) to (3.10).

$$Y = [(1 - a_k)K]^{\alpha}[A(1 - a_L)L]^{1-\alpha} \tag{3.7}$$

$$\dot{A} = B(a_k K)^{\beta}(a_L L)^{\gamma} A^{\vartheta} \tag{3.8}$$

$$\dot{K} = sY \tag{3.9}$$

$$\dot{L} = nL \tag{3.10}$$

Where a_K and a_L are the fractions of capital and labour used in the research and development sector (R&D), $(1 - a_K)$ and $(1 - a_L)$ are the fractions used in the goods producing sector, $0 < \alpha < 1$, $B > 0$, $\beta \geqslant 0$, $\gamma \geqslant 1$, and a dot over a variable denotes its derivative with respect to time.[27]

For simplicity we eliminate capital from the model

$$Y = A(1 - a_L)L \tag{3.11}$$

$$\dot{A} = B(a_L L)^{\gamma} A^{\vartheta} \tag{3.12}$$

Since Equation (3.11) implies that output per worker y is proportional to A, the growth rate of y is the same as the growth rate of A, which is given by

$$g_A = \frac{\dot{A}}{A} = Ba_L^{\gamma} L^{\gamma} A^{\vartheta - 1} \tag{3.13}$$

From Equation (3.13) we obtain

$$\dot{g}_A = [\gamma n + (\vartheta - 1) g_A] g_A \tag{3.14}$$

g_A is constant when $\gamma n + (\vartheta - 1) g_A = 0$. Therefore $g_A^* = \gamma n/(1 - \vartheta)$.

The long-run growth rate of output per worker g_A^* is an increasing function of the rate of population growth n.

R&D models of growth tend to show a positive association between population growth and the growth rate of the economy. David Romer (1996) in commenting on this result points out the fact that this seems to be counterfactual, since the growth rate of output per worker is not on average higher in countries with faster population growth, and suggests that the correct interpretation may be that higher world-wide population growth raises world-wide income growth.

In his 1990 model of R&D and growth – a model which represents a major advance of growth theory as discoveries depend in part on purposive research, development effort and innovations spread only gradually to other producers and hence imperfect competition arises – Paul Romer states that the effect of an increase in L on the rate of growth is ambiguous, something that a priori theorising cannot resolve (Romer 1990: 95).

In terms of the simplified model of R&D and growth described by Equations (3.7) to (3.10), the Romer (1990) model could be represented as a version with $\beta = 0$ (the research sector uses no capital), $\theta = 1$ (constant returns to knowledge in knowledge production) and $\gamma = 1$. In this exemplification the rate of growth of A would be proportional to L, $\dot{A} = Ba_L LA$, and the model is identical to the Solow model with $n = \delta = 0$ and with the rate of technological progress equal to $Ba_L L$. Since there is no population growth, the growth rate of output and capital on the balanced growth path would be $Ba_L L$ (Romer 1996).

Yet, in the actual 1990 P. Romer model the economy has three sectors. The research sector uses human capital and the existing stock of knowledge to produce designs for new producer durables; an intermediate goods sector uses the designs from the research sector together with forgone output to produce producer durables for use in final goods production at any time; a final goods sector uses labour, human capital and the set of producer durables that are available to produce final output. Human capital is a measure of the cumulative effect of person specific formal education and on-the-job training. Total human capital H is given and it is devoted either to the final output sector (H_y) or to the research sector (H_a). The assumption is that anyone engaged in research has free access to the entire stock of knowledge. The output of researcher j is therefore $\delta H_j A$ where δ is a productivity parameter and the aggregate stock of designs – the variable A – evolves according to $\dot{A} = \delta H_a A$ where H_a is total human capital employed in research.

When the model is solved for the balanced growth equilibrium, the common growth rate for all the variables is δH_a, i.e. it is proportional to the amount of human capital employed in the research sector. Human capital is the relevant scale variable because it is the input that is used most intensively in research. The parameter L is not among the determinants of the growth rate because the aggregate supply of L and H is considered fixed and the fraction of human capital devoted to research does not increase when L increases.[28] The effects of increases of L and H that have been observed historically are left out of the analysis.

An author who contended that the long-run history of population growth and technological change is consistent with the population implications of endogenous technical change, i.e. with the result that high population spurs technological change, is Kremer (1993).

The model, which is similar to Lee (1988), uses two basic notions. First, the idea – already advanced by Kuznets (1960) and Simon (1977, 1981) – that a higher population means more potential inventors; second, the Malthusian assumption that population is limited by the available technology.

A simple version of the model is described by Equations (3.15) to (3.18).

$$Y = Ap^{\alpha}T^{(1-\alpha)} \tag{3.15}$$

Where A is level of technology, p is population and T is land.

When T is normalised to 1, per capita income y can be written as $y = Ap^{(\alpha-1)}$, therefore, assuming that population adjusts so that output per person equals the subsistence level $\bar{y}$, we obtain

$$p = \left(\frac{\bar{y}}{A}\right)^{1/\alpha - 1} \tag{3.16}$$

The next equation describes the relationship between technological change and population, assuming that the growth rate of technology is proportional to total population, because population determines the number of potential inventors.

$$\frac{\dot{A}}{A} = pg \tag{3.17}$$

where g represents research productivity per person.

Taking the logarithm of Equation (3.16) and differentiating it with respect to time yields

$$\frac{\dot{p}}{p} = \frac{1}{1 - \alpha} \frac{\dot{A}}{A} \tag{3.18}$$

and, considering Equation (3.17) we obtain

$$\frac{\dot{p}}{p} = \frac{g}{1 - \alpha} p \tag{3.19}$$

In this model the growth rate of population is proportional to the growth rate of technology and this in turn is proportional to the level of population: hence the result that the growth rate of population is proportional to the level of population. This conclusion, according to Kremer, matches the pattern of world population growth over most of history, and also accords with the fact that among societies with no opportunity for technological contact those with greater initial population attained higher technology levels and population densities.[29]

In the model just discussed population is made endogenous through its relation with per capita income. A similar possibility is explored by Nerlove and Raut (1997) who introduce endogenous population within the Solow model – by supposing simply that the rate of growth of population depends on the level of per capita consumption. This introduction entails the possibility of multiple equilibria – characterised by large populations with low per capita incomes or by small populations with high per capita incomes – and instability of some of the equilibria. The authors suggest the conclusion that 'without knowing how population change is related to the stock of capital, the level of population and other state variables of the economy . . . there is little to distinguish which path may be followed . . . It is clear that merely endogenizing population growth at the macro level does not shed light on . . . the nature of the dynamics' (Nerlove and Raut 1997: 1171).

Microeconomic analysis of families' choices

The relation between economic and demographic change is explored at a different level by the literature on the economics of the family, which mostly uses a neoclassical framework to provide microeconomic analysis of families' choices and especially of the motives for limiting family size. Non-market activities of household are viewed as optimising decisions representing economic choices and changes in the family are viewed as a result of external shifts in prices and incomes caused by technological change.[30] In the 'household demand' model of fertility families maximise a utility function depending on children quantity (N), children quality (Q), and other consumption goods (C):

$$U = f(N, Q, C) \tag{3.20}$$

subject to a production function constraint in which the production of children and of C requires inputs not only of conventional goods but also of time.

Although children are normal goods – i.e. their demand increases with income – high and rising income may be associated with low and falling fertility. Since each child imposes a time cost of rearing, as the value of time (its opportunity cost in terms of wage) – and especially the value of women's time – increases, the quantity of children becomes more expensive. As Schultz (1973) points out 'Whereas Malthus assumed that the price of children would remain constant . . . the cost of children increases with the rise in price of human time'. An increase in wages induces households to choose small families and larger capital (physical and human) investments on each child (Birdsall 1988; Tamura 1995).

A growth model in which economic development influences family decisions about the number of children and, hence, the fertility rate, is presented by Barro and Sala-i-Martin (1995). The model uses (in a continuous time framework) the results of Becker and Barro (1988) that relate parents' utility to the utility of their children through the function

$$U_0 = u(c_0, n_0) + a(n_0)n_0U_1 \tag{3.21}$$

where U_0 is the utility of an adult, U_1 is the utility of each child, n_0 is the number of children per adult, and $a(n_0)$ represents the degree of altruism towards each child. Becker and Barro assume that $a(\cdot) > 0$, $a'(\cdot) < 0$, i.e. the utility of an adult is increasing in the utility of each child (one-sided altruism), but the utility of an additional child is decreasing in the number of children. Hence the utility of an adult is increasing and linear in the utility of each child and increasing and concave in the number of children.[31]

As for the family's budget constraint, a key assumption is that a child's rearing cost is $\eta = b_o + b_k$ where b_o represents the goods cost of child rearing and the b_k part represents a part of the cost that increases with the capital intensity k, which in the model represents the per capita quantities of human and physical capital. This formulation reflects the notion that, because child rearing is intensive in parental time, its cost tends to rise with parents' wage rates or with other measures of the opportunity costs of such time. Hence, the family's budget constraint in the model is expressed as

$$\dot{k} = w + (r - n + d)k - bnk - c \tag{3.22}$$

where $\dot{k}$ is the growth rate of the household's own per capita assets, w and r are respectively the wage rate and the rate of return on assets, n and d are the family's birth and mortality rate, c and k are the family's per capita consumption and assets respectively.

The transitional dynamics of this model imply that if the economy begins with a capital–labour ratio lower than its steady-state values, k increases and c/k – which expresses the ratio of the income effect on the demand for children, represented by c, to the cost of children, which depends linearly on k – declines monotonically towards its steady-state value. Since during the transition to the steady state the fertility rate moves in the same direction as c/k, n also declines. That is (with a given mortality rate) fertility declines steadily as the economy develops, a result that accords with empirical evidence. The positive association between fertility and per capita GDP which the data show at extremely low levels of income appear when considering the goods cost $-b_o$ of child rearing, as this introduces a positive income effect in the relation between fertility and per capita product. Moreover, since the goods cost is relatively more important in poor countries, the net positive relation between fertility and per capita product tends to appear only at low levels of per capita product.

Becker *et al.* (1990) obtain more complex results in terms of the population-development linkages in a model with endogenous fertility. The model uses the Becker and Barro (1988) dynastic utility function and two additional assumptions. An increase in the stock of human capital raises per capita income and hence has a positive income effect as well as a negative substitution effect – the cost of the time input must rise as *H* increases – on the demand for children. The income effect dominates in economies with little human capital but the substitution effect eventually becomes dominant when *H* is large enough. The additional, key assumption is that the rate of return on human capital *H*, unlike for physical capital, does not monotonically decline as the stock of human capital increases. Rates of return are low when there is little human capital and they grow at least for a while as human capital increases.

The dynamics of the model are described by Figure 3.4.

The function *hh* gives human capital in period $t+1$ as a function of the amount in *t*. Low levels of H_t generate $H_{t+1} < H_t$ and high levels of H_t generate $H_{t+1} > H_t$.

The shape of the *hh* function depends on the behaviour of the rates of return on investments in human capital versus the discount rate on

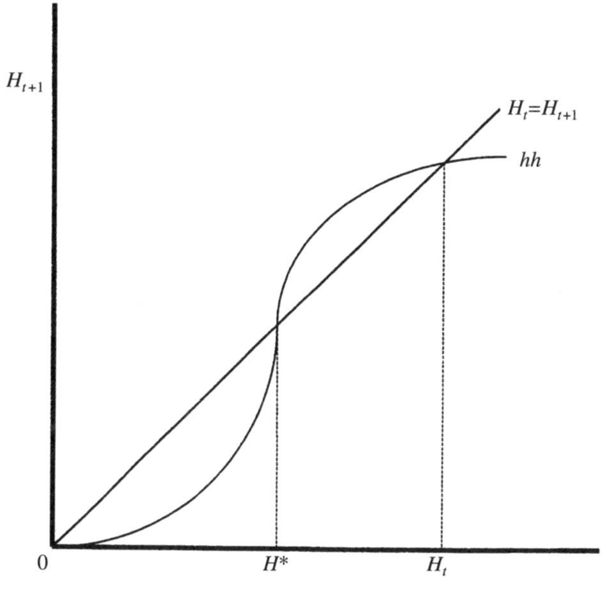

Figure 3.4 Human capital and growth.

Source: Becker *et al.* (1990)

future consumption. At the origin and at low levels of H the economy does not invest in human capital or the amount invested is less than the capital that wears out (because low H implies high fertility which in turn raises the discount rate on the future while the rates of return on H are low). Thus there is a locally stable steady state at $H = 0$. Countries may be trapped in a Malthusian, underdeveloped steady state, with high birth rates and low level of human capital. As the amount invested in H continues to rise, fertility declines, the discount rate on future consumption decreases while the rate of return on H increases and a new steady state emerges (when H is sufficiently large that the rate of return equals the discount rate).

In summary, it appears that recent analytical models tend to generate a plethora of conclusions about the relation between economic and demographic variables. Hence, neoclassical growth theory – beyond Solow – offers many insights but no simple wisdom or common ground to the analysis of the population–economic development nexus.

Part II

Population growth and agriculture in less developed countries

4 Endogenous technical progress in agriculture

Trends

The basis of Malthus pessimism was the idea that population growth would translate into increasing pressure on land and therefore, in the absence of technical progress, into decreasing returns to labour. If population growth is associated with decreasing land to labour ratios, a possible pessimistic scenario – outlined by Khan (1991), who refers to the experience of Bangladesh in the 1950s – entails not just lower agricultural incomes, but also landlessness. Incidentally it must be remarked that, with a given land endowment, landlessness will more likely result from population growth if the pattern of land ownership is very unequal. The resulting wage–labour supply increase and the increased factor share of land then translate into lower real wages. A similar situation is described by El-Ghonemy (1990) for Egypt during 1897–1950, when the agricultural population doubled: the man–land ratio, landlessness, seasonal unemployment and child labour utilisation increased substantially as a consequence, and labour productivity in agriculture declined steadily.

Population growth may not imply the lowering of land to labour ratios if the land frontier can be expanded or if labour absorption in the non-agricultural sector is fast (or substantial migration outlets exist, as was the case for modern European growth). Neither the first option nor the last, however, are available to most of today's developing countries.

In the developing countries the land used in crop production is only a fraction of the total land with rain-fed crop production potential – about 30 per cent according to FAO (Alexandratos 1995a). However more than 90 per cent of the unused land is concentrated in sub-Saharan Africa and Latin America/Caribbean – while there is little scope for expansion in other developing areas – almost half of this

Table 4.1 Growth rates of population, aggregate demand and production (per cent p.a.)

	1967–97			*1977–97*			*1987–97*		
	Pop.	*Dem.*	*Prod.*	*Pop.*	*Dem.*	*Prod.*	*Pop.*	*Dem.*	*Prod.*
World	1.7	2.2	2.2	1.6	2.1	2.1	1.5	1.8	1.9
Developing countries	2.1	3.6	3.4	2.0	3.7	3.6	1.8	3.9	3.8
excluding China	2.3	3.2	2.9	2.2	3.1	2.9	2.1	3.0	2.9
sub-Saharan Africa	2.8	2.7	2.2	2.9	2.9	2.7	2.7	3.2	3.0
Latin America and Caribbean	2.1	2.9	2.8	1.9	2.5	2.4	1.8	2.6	2.5
Industrial countries	0.7	1.1	1.3	0.7	1.0	1.0	0.7	0.8	1.3
Transition economies	0.7	0.3	0.1	0.6	−1.1	−1.0	0.2	−4.2	−4.1

Source: FAO (2000)

land is under forest and a significant part of it is likely to suffer from soil and terrain constraints (Alexandratos 1995a).

As for labour absorption, this has been fast in the newly industrialised countries, but most developing countries have seen increasing population pressure on land. Indeed, as the World Bank (1984) points out, the high growth rates of population (and of the labour force) in today's developing countries imply growth of the absolute dimension of the agricultural labour force in spite of structural transformation – i.e. in spite of the decreasing share of agriculture in the labour force and in total product.

Yet, as Table 4.1 shows, in contrast with the Malthusian prediction, total agricultural production has more than kept pace with population. Therefore per capita production has increased during the last three decades, on a world scale and in all developing regions with the exception of sub-Saharan Africa.[32]

Moreover, such growth has basically occurred in the context of a long-term decline in the real price of food, suggesting that agriculture could have responded to faster growth of effective demand and that the persistence of undernourishment in the developing countries is not explainable with the limitations imposed by the global productive potential of agriculture (FAO 2000).[33] Moreover, projections of future supply–demand balances – obtained with different models – tend to show that 'the world as a whole, could continue to increase agricultural production at a rate faster than that of population growth and in any case at a rate that could match the growth of effective demand, probably at non increasing prices' (Alexandratos 1995b: 29).

These trends could be taken as *prima facie* evidence of the ability of agriculture to respond to increasing population without incurring Malthusian outcomes – and they are the result of the variable which was missing in Malthus's reasoning i.e. technical progress.

At the same time however these global trends do not exclude the possibility of severe local failures nor do they demonstrate the ability of agriculture to match the growth of demand without harming the environment and its own resource base.

The following sections and Chapter 5 examine the models that explain the 'success' of agriculture in responding to population pressure, while the reasons and evidence for failures are the subject of Chapters 6 and 7.

The Boserup model

In her 1965 essay *The Conditions of Agricultural Growth* and in subsequent work Ester Boserup has provided convincing reasoning and rich evidence to support a model of endogenous technical progress in agriculture that predicts intensification in response to increased population pressure on land. The starting point of Boserup's reasoning is a radical departure from the view of agriculture that was common to Malthus and Ricardo.

Agriculture and land fertility play a central role in Malthus's and Ricardo's reasoning about economic growth. As discussed in Chapter 2, in Malthus's model there are decreasing returns to labour with fixed land endowment and population is endogenous (the birth and death rates vary in response to changes in the standard of living). When population grows the real wage decreases because food production growth is less than proportional and, as a consequence, the birth rate declines and the death rate increases, until the original equilibrium is restored. Food production therefore establishes the limits to population growth.

Ricardo's theory of rent is based on the notion that there are lands with different fertility and the better lands are cultivated first: cultivation therefore can only expand into successively less productive land (the extensive margin) or into lands increasingly distant; in addition, the application of successive doses of capital or labour to existing fields (the intensive margin) would yield diminishing returns. Since profits on the better land are higher (given a uniform rate of wages) owners can obtain from farmers a rent for its use, the value of rent being such that it leaves a uniform rate of profit, independently of the quality of land used. As land of poorer quality is cultivated, a rent arises on the intermediate land, and the rent on the better land rises, until there is a lower, uniform rate of profit.

Hence, Ricardo's view of agriculture and land fertility leads him to the conclusion that the natural tendency of profits is to fall, and the economy will tend to a stationary state. The same view of land fertility had led Malthus to predict an equilibrium with stationary population and real wage. Such a view, which is based on the distinction between permanently cultivated fields and uncultivated land, is rejected by Boserup as being a generalisation of the specific circumstances which could be observed in nineteenth-century Europe, that is a distinction between permanently cultivated land (in Europe) and the expansion into new land by European settlers in the new continents. This distinction, according to Boserup, is an over-simplification of agricultural expansion: when the conditions of tropical agriculture are considered, one can observe that, across regions, low population density is associated with long fallow and higher population density with a progressive shortening of fallow, until annual cropping or multicropping prevail; in a broad historical setting, the development of agriculture has occurred through a gradual shortening of fallow, with associated changes in technique.

Substituting the distinction between fields and uncultivated land with the frequency of cropping, as in Boserup, has profound implications for the relationship between population pressure on land and land productivity. While in Malthus's model, population is endogenous (the birth and death rates vary in response to changes in the standard of living) and there is no significant technical progress in agriculture, in Boserup land fertility becomes a variable positively associated with population density and related changes in system of land use and techniques. Population is exogenous and technical progress in agriculture, endogenous.

Boserup classifies the systems of land use in categories based on the length of fallow, each category being typically associated with the use of a different set of tools, as shown in Table 4.2. This sequence can describe both the main stages of the evolution of agriculture from prehistoric times, and the association between different population densities and systems of land use in different countries.

Table 4.2 Food supply systems

Food supply systems	*Description*	*Tools used*
Gathering	Wild plants, roots, fruits, nuts	N/A
Forest fallow	One or two crops followed by 15–25 years of fallow	Axe, machete, digging stick
Bush fallow	Two or more crops followed by 8–10 years of fallow	Axe, machete, digging stick, hoe
Short fallow	One or two crops followed by one year or two years of fallow	Hoes, animal traction
Annual cropping	One crop each year	Animal traction, tractors

The essence of Boserup's explanation of the link between population growth and change in systems of land use and technology is outlined in Figure 4.1.

Figure 4.1 refers to a community that can use a given amount of land and where population growth results in an increased number of workers per acre (L). As L increases, the community will move along production function F_1. Yield per acre increases but at a diminishing rate, because the land factor is fixed, and average output per worker, as shown in the upper left quadrant, declines along AP_1. F_2 represents another technology, associated with a different system of land use. This technology is inferior to F_1 (i.e. it implies lower yields and lower average product) at low values of L.

For example, with the system of shifting cultivation in forests, land clearing is easily done through fire, and there is no need for labour intensive tasks such as land preparation, weeding and manuring, and care of draught animals. Long fallow is a labour saving way to restore fertility. With the shortening of fallow, less land has to be cleared in a given year, but the reduction of labour requirements is outweighed by the additional labour necessary for preparation, manuring and weeding. Hence, at very low densities, shifting cultivation in forests with the use of fire is associated with higher yield per acre and output per worker than would be the case with more intensive methods. In this setting however only a small fraction of the land available to a community can be cultivated in a given year.

As population grows, the community moves along a diminishing returns path, until eventually it becomes convenient to switch to a new

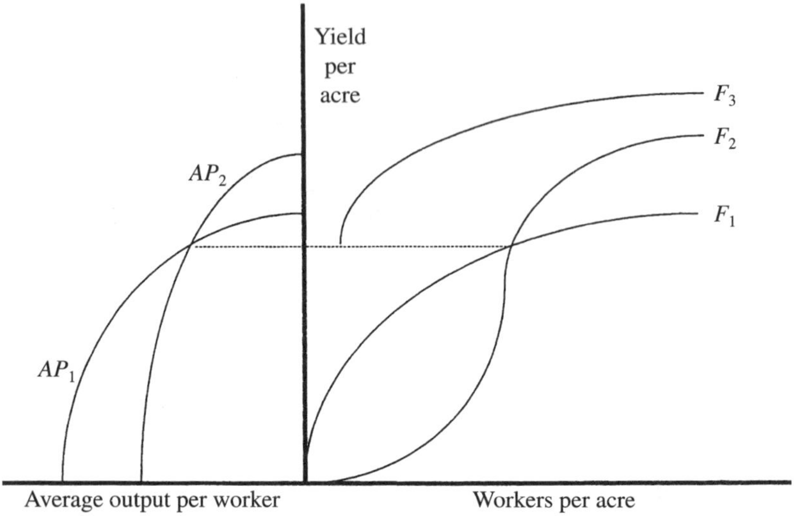

Figure 4.1 The Boserup model.

Source: adapted from Robinson and Schutjer (1984)

system of land use and technique (F_2). Yields per acre increase, but output per hour worked may decline or stagnate. In discussing the introduction of animal-drawn ploughs, for instance, Boserup comments that a new technology is to be regarded as a means to prevent a fall in output per man hour, rather than a means to raise it (Boserup, 1965: 34). She is more generally inclined to think that a period of sustained population growth will result in a decline of agricultural labour productivity, although, in the long run, the positive effects of population growth in terms of general economic development may eventually lead to increased agricultural labour productivity (Boserup 1965: 118).

We have therefore a model where growing pressure on land implies increasing use of labour with associated diminishing returns until, eventually, a new, superior technique is introduced. The process occurs in response to increasing food needs within communities and allows for the satisfaction of such needs. Boserup's horizon is very broad, describing a typical sequence of agricultural development from prehistoric times to the recent past in which land use systems and techniques evolve in response to increasing population pressure. Optimism, as we have seen, is actually confined to food production, since labour productivity may decline or stagnate in the process (Boserup 1965, 1981; Robinson and Schutjer 1984).

The model explains the transmission, rather than the invention, of technologies in low industrial input agriculture. In *Population and Technology* (1981) Boserup states that it is more important to focus on the conditions for transmission of techniques than on those for the appearance of inventions, since societies have more often advanced by introducing technologies already known in other societies, and in *The Conditions of Agricultural Growth* (1965) she argues that in economies at low levels of industrialisation most private investment in agriculture is done by local labour with traditional tools (1965: 102).

In this sense, as Hayami and Ruttan (1985) point out, Boserup shares with the classical economists a perspective of agriculture as an essentially self-contained system, although in her case this is the result of focusing attention on less developed countries. In Figure 4.1 technologies F_1 and F_2 are adopted respectively if labour is scarce, and when land becomes scarce, in both cases with no use of industrial inputs: indeed such technologies are best understood as being associated with two different lengths of fallow periods. However, at a more developed stage, both labour and land are effectively substituted with capital: technology F_3 in Figure 4.1 is superior for all densities.

The shift to high input agriculture and the issue of the relationship between population and the emergence of a significant agricultural surplus is not really explained by Boserup, although in some of her writings there are interesting insights on this issue, basically suggesting that population growth is linked to the emergence of commercial, high input agriculture through its links with urbanisation.[34]

It is necessary to underline the fact that, since the model in essence refers to an egalitarian subsistence economy, the maintenance of per capita food production is equivalent to the maintenance of per capita consumption levels, a feature that is obviously very simplistic and can be misleading. In the context of more complex economies – with a non-agricultural sector, differing resource bases, unequal distribution of income and assets and international trade – such equivalence does not apply, since, for instance, food can be imported. Furthermore, as Sen has illustrated, adequate per capita food production growth does not guarantee that each individual or social group can avoid starvation.

Nevertheless, before discussing more extensively the limitations of the model, one must recall that there are critical links between the satisfaction of individual food needs and food production: global food production is relevant, because of the price of food; countries' own per capita agricultural production is relevant, especially when their capability to import through non-agricultural exports is limited and when agriculture's share in the economy is high, and therefore its growth performance crucially affects income and effective demand (Sen 1981, 1994; Alexandratos, 1995a, b).

A general idea of the relevance of Boserup's analysis for contemporary developing countries can be conveyed by data on labour–land ratios and on land and labour productivity. Figures 4.2 and 4.3 show, for 1962–92, the percentage changes in the labour–land ratio (*x*-axis) as compared to the percentage changes in land productivity (Figure 4.2) and in labour productivity (Figure 4.3) (*y*-axis).[35]

Since the majority of countries in Figure 4.2 fall in the upper right quadrant, it appears that population pressure on land has increased in most cases and a Boserupian, land augmenting response has also occurred with few exceptions. The cases which belong to the left quadrant are mostly countries which have experienced rapid industrial development and countries in oil producing regions, where there has been a significant absorption of labour in the non-agricultural sectors. In some cases, notably Brazil, an important role has also been played by a great expansion of agricultural land. The countries which show increased population pressure and stagnant yields have the common characteristic of having suffered harsh and prolonged civil strife (Angola, Ethiopia and Mozambique).

Labour productivity growth (Figure 4.3), however, has in general been far less substantial, and in many cases productivity has hardly improved in 30 years, or has declined. The majority of such cases are African countries where the yield increases have been on a par or below the variations of the land–labour ratios (countries lying around or below the 45 degree line).

The general conclusion that can be derived from these highly aggregated data is that, as Boserup postulated, although land intensification tends to take place in association with population growth, optimism does not extend to labour productivity, since increases here have been generally lower than for yields and many countries have experienced a decline.

Several country studies (Pingali *et al.* 1987; Pingali and Binswanger, 1987; Turner *et al.* 1993; Tyffen 1995) have confirmed Boserup's main arguments[36] and it is difficult to disagree with Lipton's (1990: 223) observation that the Boserup model, unlike that of Malthus, may offer an explanation for the difference in vital statistics and agricultural

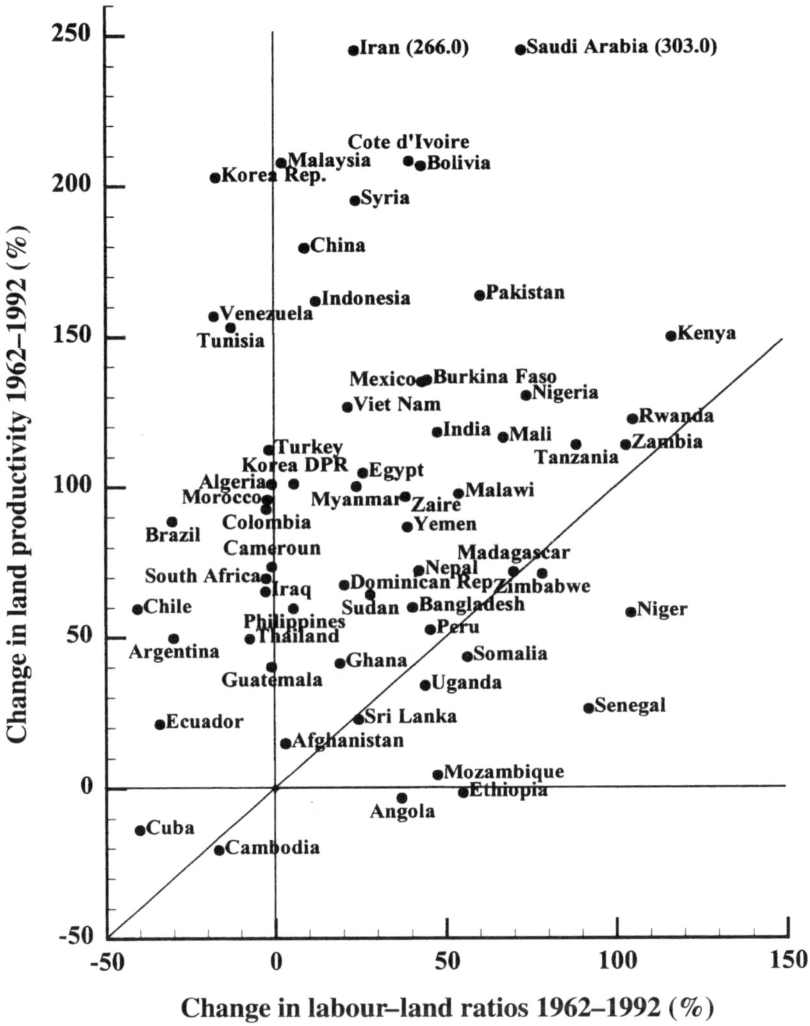

Figure 4.2 Population pressure and land productivity.

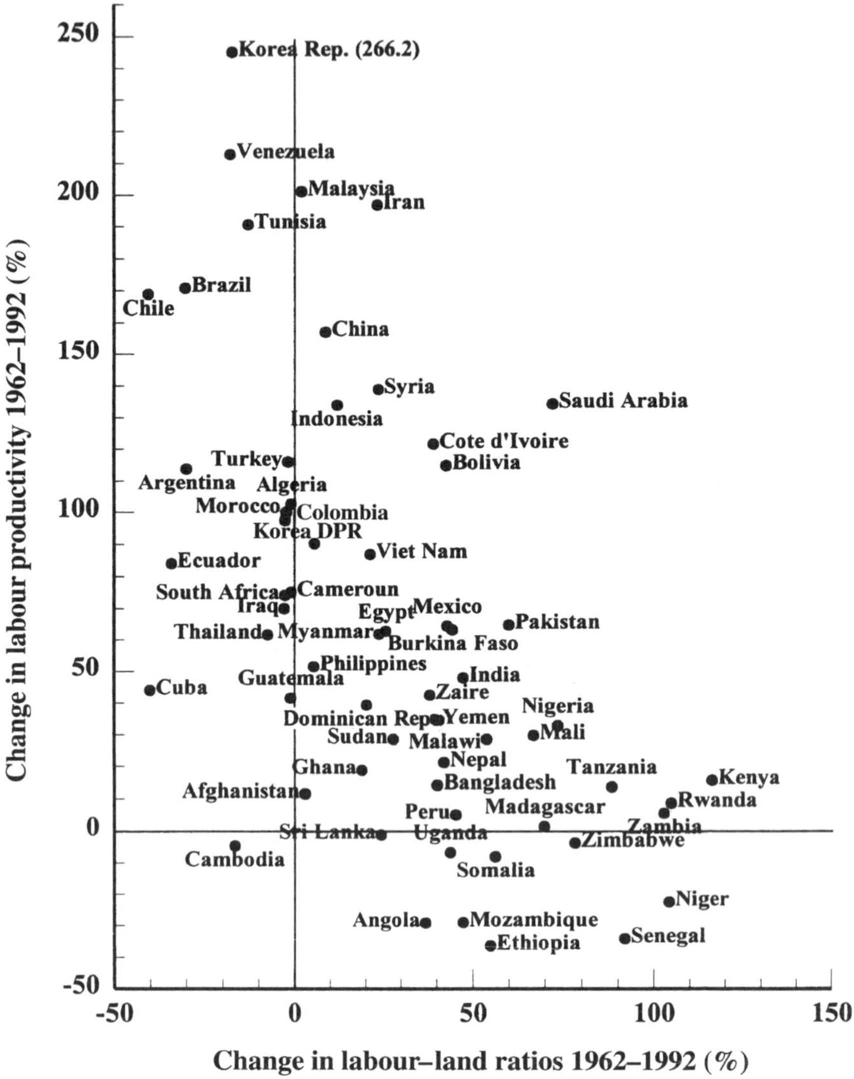

Figure 4.3 Population pressure and labour productivity.

growth performance between land-abundant Africa and South and East Asia.

When using cross-sectional data on population and land productivity one should consider the possible linkages running in the reverse direction, i.e. from agriculture to population. The classical Malthusian linkages are that improved nutrition could lead to higher birth rates and lower mortality. In addition, differences in yield per person could be expected to result in migrations from low to high-yield areas. The importance of migrations in this context obviously tends to be much more relevant within countries than it is between countries, and should therefore have a limited effect on cross-country comparisons.

As for the Malthusian linkages in the LDCs, their pattern seems to accord with the general pattern of the relationship between income growth and population forecast both by transition theory and by the household economics models discussed in Chapter 3 (pp. 59–62). In the short run the Malthusian linkages may operate in contemporary LDCs' agriculture. Lipton (1997) points out the fact that the Indian districts that achieved more technical progress in agriculture in 1960/1–1987/8 (almost all during the green revolution) experienced slower declines in 1971–81 in total fertility rates, and hence faster population growth, than the districts with less technical progress along green revolution lines. However, in the long run the Malthusian fertility responses to agricultural growth tend to fade, since income growth results in a limitation in the size of the family. Therefore, long-run cross-sectional associations between population and agricultural growth should be expected to reflect a positive linkage running from population to agriculture.[37]

Many theoretical shortcomings limit the validity of the Boserup model. First, the egalitarian structure of the model, whereby increasing food needs induce technological responses and increased food production caters to the needs of the community, overlooks all the questions of distribution and entitlement. Addressing the link between necessity and invention, Lipton (1990) has argued that necessity must not be hunger alone, but rather hunger backed by effective demand, and that in some very poor, intensive-agriculture areas of Asia with severe income inequity, this is not the case. In addition, as Sen (1981) has extensively argued, adequate food production growth does not guarantee that each individual or social group has enough access to food to avoid starvation. Command over food depends on what a person owns (land, labour power and other resources) and on how this endowment enables him or her to obtain food.[38] Thus, if population growth is associated with increasing landlessness, technical change must not only make it possible to increase land productivity: it must also be labour-using, or an increasing number of unemployed wage workers may be exposed to starvation.

Second, the Boserup model focuses on technology transmission – it does not explain the entire process of technical change, which includes research, discovery and adoption – therefore it must be assumed that at some point in space and time, necessity has mothered invention, and that such invention can be adopted elsewhere. One should, however, provide an economic explanation of invention; in addition, agricultural technologies, especially land augmenting ones, are often location-specific and transmission is therefore not simple.

A third, related problem, is that of institutions: do institutions (in this case the research system and property rights on land) adapt, and, if so, how and at what pace? Fourth, as already discussed, the model focuses on low-input agriculture, and therefore neglects the fact that limited availability of capital to acquire inputs may be a constraint to technology adoption. Furthermore, certainly realism and insight could be gained by considering intensification not just in terms of yield increases as in Boserup, but also in terms of shifts from low- to high-value crops on any given land and of geographical shifts in crop production from areas of low potential to those of higher potential, as suggested by Lele and Stone (1989).

Finally, and perhaps crucially, the model is based on historical data and slow processes of adaptation and neglects some possible critical implications of the rapidity of changes and/or the already high densities in contemporary poor countries, implications which are discussed in Chapters 6 and 7.

The induced innovation model

Optimism over the questions left open by Boserup is suggested by the Hayami and Ruttan (1985) induced innovation model (IIM). This model provides an economic theory of invention and adoption of technologies, whereby relative price changes determine responses by economic agents and research institutions, that guide technological progress towards saving the factors that become relatively scarce. Thus, as population grows, land prices increase relative to wages and to the prices of man-made, land-saving inputs (such as fertilisers). Technological change therefore moves in a land saving, labour using direction. Moreover, Hayami and Ruttan argue that changes in underlying economic conditions (increasing population pressure on land, increasing product demand because of trade opportunities, and technical change) also induce the development of property rights and contractual arrangements that allow a more efficient allocation of resources.[39]

From the point of view of the general approach to agricultural development, the IIM essentially belongs to what Hayami and Ruttan have defined the 'high-payoff input' model of agricultural development of T. Schultz (1973). During the 1960s Schultz had presented strong arguments against the notion that traditional farmers were inherently incapable of responding to economic incentives. In his book *Transforming Traditional Agriculture* he argues that the backwardness of traditional agriculture is not due to the 'irrationality' of farmers in developing countries, but rather to the fact that agricultural technology is location specific and therefore the transferability of techniques from developed countries is limited. Hence, agricultural development would require, essentially, investment in research and technical assistance. The success of this representation of the problem is strongly related to the contemporary success of what has been termed the 'green revolution', that is the introduction of high-yielding varieties of grains in several parts of the developing world from the mid-1960s. The Hayami and Ruttan model offers considerable microeconomic and empirical support to this approach, while explaining the economic and institutional mechanisms that determine the direction of technical progress.

The IIM moves from the observation that countries with differing factor endowments have followed different paths of technical change in agriculture and provides an explanation of such diversity that belongs to the debate on innovation biases, and specifically refers to the Hicksian notion of induced bias of innovation.

In *The Theory of Wages* Hicks (1966) suggested a classification of inventions as 'labour saving', 'neutral' and 'capital saving' and introduced the notion of 'induced' inventions, in the sense of induced bias, in the context of a discussion of the relationship between distribution and growth. According to his original definitions, an invention is adopted only if it achieves the same product with a smaller amount of resources; inventions are labour saving if their initial effect is to increase the ratio of the marginal product of capital to that of labour, neutral if they leave such ratio unchanged and capital saving if they diminish it. Following this definition, technical progress can be represented as a shift in the unit isoquant closer to the origin.

The difference between Figures 4.4a, 4.4b and 4.4c is that in Figure 4.4a the technical rate of substitution, i.e. the relationship between the marginal productivity of labour (MPL) and the marginal productivity of capital (MPK) along ray R (that is, given factor proportions) is

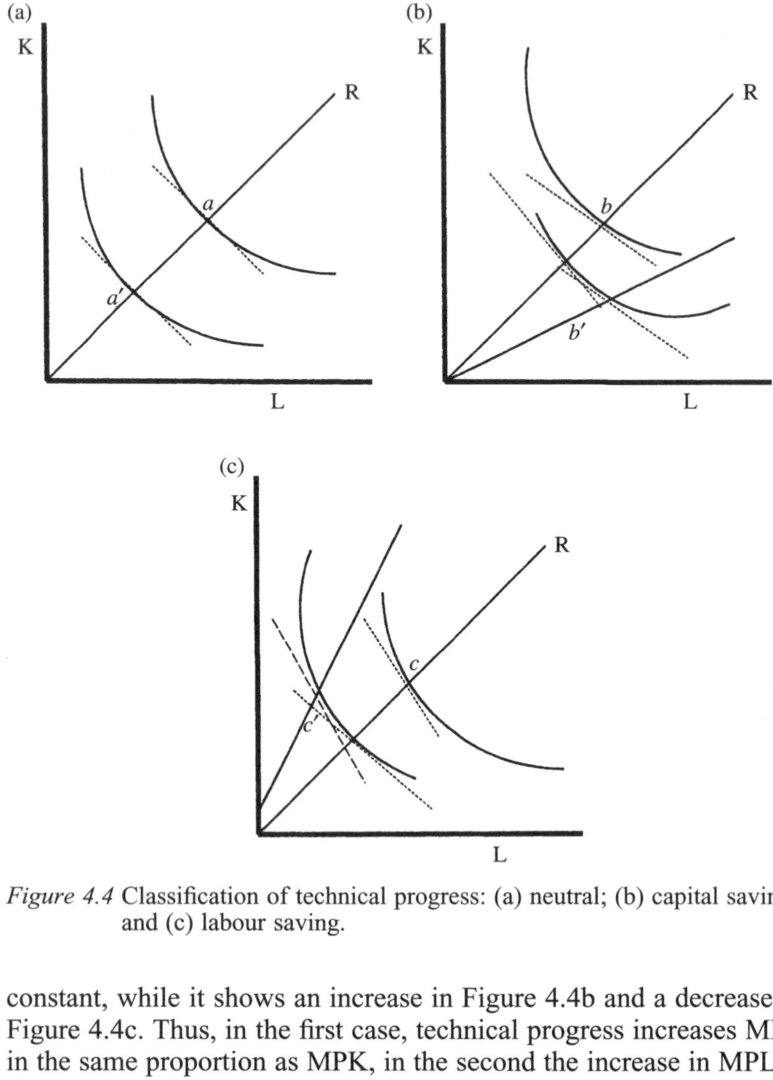

Figure 4.4 Classification of technical progress: (a) neutral; (b) capital saving; and (c) labour saving.

constant, while it shows an increase in Figure 4.4b and a decrease in Figure 4.4c. Thus, in the first case, technical progress increases MPL in the same proportion as MPK, in the second the increase in MPL is larger than the increase in MPK, in the third case the reverse is true. At constant factor prices, in Figure 4.4a there is no change in factor proportions, in Figure 4.4b the optimal combination moves to b', and in Figure 4.4c to c'. Technical progress is neutral in Figure 4.4a, capital saving, or labour using, in Figure 4.4b and labour saving in Figure 4.4c. One should note that in an absolute sense the switch to the new technique might save both capital and labour, and that the definition refers to changes in factor proportions. According to Hicks (1966: 124), 'A change in the relative prices of the factors of production is itself a spur to invention, and to invention of a particular kind, directed to economising the use of a factor which has become relatively expensive'. Hence he argued that there is no inherent labour saving bias in technical change: the introduction of labour saving innovations is a consequence of rising wages.

In the framework of this notion of induced bias of technical change, the IIM uses the concept of an innovation possibility curve (IPC), as developed by Syed Ahmad (1966). At a given time there exists a set of production processes which could be potentially 'invented' given the state of scientific knowledge. Each process can be represented by an isoquant with a relatively small elasticity of substitution, while the IPC can be represented as the envelope of such isoquants. Thus the IPC is a purely technological matter, and it might be assumed that, because of advances of basic science, it moves closer to the origin over time, and that such shift is neutral, i.e. there is no bias in the historical innovation possibilities. The prevailing price ratio determines instead the choice of a particular isoquant (representing a specific production function).

In Figure 4.5 IPC_t is the innovation possibility curve at time t and, given the price ratio PP, the optimal choice is the cost minimising isoquant I_t. At time $t+1$ the new innovation possibility curve is IPC_{t+1}. If the same price ratio prevailed the specific invention chosen would be I_{t+1}, which implies unchanged factor proportion (in this sense the IPC shift is neutral). However if the prevailing price ratio is $P'P'$ the technique I'_{t+1} is invented, resulting in a relatively lower use of the factor (labour) which becomes more expensive.

Finally, Hayami and Ruttan classify agricultural innovations as being either labour saving or land saving. Mechanical technology is labour saving, in the sense that historically the substitution of land and capital for labour in relatively labour scarce economies has been made possible primarily by the progress of mechanisation. Biological and chemical technology is land saving, in the sense that historically the dominant factor for substituting fertiliser and/or industrial inputs for land in land scarce countries has been the development of (high yielding) crop varieties.

The model of induced technical change in agriculture incorporating the concepts introduced above is depicted in Figure 4.6. Figure 4.6a refers to the invention of mechanical innovations, represented as the switch from technology I_t to technology I'_{t+1} (using relatively less labour and relatively more land-cum-power), as a response to an

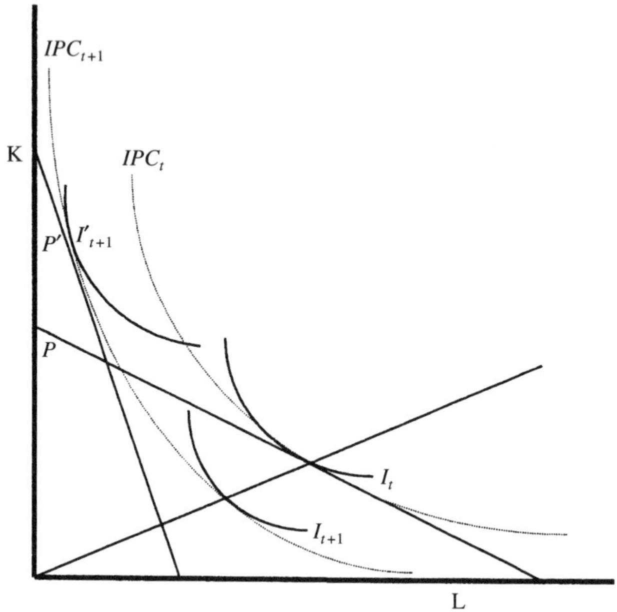

Figure 4.5 Technical progress and the innovation possibility curve.

increase in the relative price of labour (from *PP* to $P'P'$) . The process described is the invention of machinery that allows a worker to cultivate a larger area. Figure 4.6b refers to the advance of biological technology, represented as the switch to a technology (I'_{t+1}) using relatively less land and more fertiliser and land infrastructure as compared to I_t, in response to a decrease in the relative price of fertiliser. A new high-yielding variety of a crop, requiring more fertiliser application and more irrigation is the type of innovation described.

The IIM is consistent with an extensive body of cross-sectional and time series observations. Hayami and Ruttan carry a comparative analysis of the history of agricultural development in Japan and the USA between 1880 and 1980, showing that the two countries have experienced similar, high rates of growth of agricultural output and of labour productivity. However, in the land abundant USA the growth of labour productivity was associated with increases in the land–labour ratio and with the introduction of mechanical innovations; in land-scarce Japan the growth of labour productivity was associated with increases in the yield–labour ratio, and with the introduction of

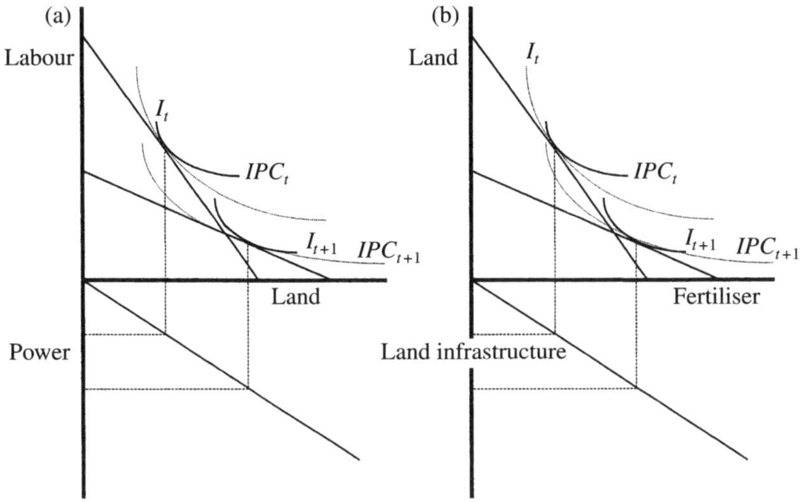

Figure 4.6 The Hayami and Ruttan induced innovations model: (a) mechanical innovations; (b) biological innovations.

Source: Hayami and Ruttan (1985)

biological innovations. Furthermore, the analysis of the growth in labour and land productivity between 1960 and 1980 in a large number of developed and developing countries shows that the lower the land–labour ratios the more labour productivity increase has been associated with high yield per acre, suggesting that innovation has been consistent with the underlying factor endowment.

In summary in this model relative price changes induce innovations that would save the factors that become relatively scarce. This is relevant to our topic for two reasons: one is that population growth might induce the type of land intensification and increasing yields that are needed to satisfy the increasing food requirements of the community. The other reason is that such a mechanism in principle should ensure that labour is not displaced.

But how can price changes influence the whole process of research and innovation? Hayami and Ruttan surmise that shifts in relative prices induce farmers to demand innovations that save the factor whose relative price increases. Both the public research system (scientists and administrators) and the private firms supplying modern inputs are supposed to respond by producing the right kind of innovation. Therefore the success of the induced innovation mechanism requires that institutional change – in this case the response of the research system – be guided by the movements of relative prices.

The theory outlined clearly belongs to neoclassical institutionalism:[40] changes in relative prices in this model determine the process of research resources allocation. The authors however recognise two major sources of failure. The first is that, in the case of agriculture, the public-good attributes of biological technology result in a less than optimal level of supply by private firms;[41] the second is that innovations will be supplied only if the expected returns to political entrepreneurs (in terms of consensus) exceed their costs for the entrepreneurs. Hence the supply of innovations is influenced by the power structure of society and may not be optimal. While this latter line of thought is not really developed, the former has been a major policy implication of the IIM, as this argument has been extensively used to argue for the need of public research in land-saving technology.

The plausibility of the induced innovation mechanism hypothesised by Hayami and Ruttan is subject to the general critiques that apply to neoclassical institutionalism, discussed in Chapter 3. Economists working within the induced innovation framework have criticised the failure of the model to fully account for sources of distortions especially relevant for the economies of less developed countries, first among all the impact of the unequal distribution of land among farmers, and the power structure of society which ensue from this.

In a 1973 article, de Janvry developed a model of induced innovations that differs from the IIM model in two ways. First, the expected payoffs from agricultural innovations differ among different social groups. Such expected payoffs determine the latent demand for innovations of each group. The extent to which these latent demands are actually conveyed to the research system depends on the capacity each group has to press its specific interests through the politic–bureaucratic structure. De Janvry holds that in the case of Argentina the actual demand for public innovation results from the maximisation of the utility function of the dominant farm interests.

The decision mechanism of the inducement of innovations by the traditional landed élite is described as follows

$$LU\{\Pr(\pi \geqslant 0) = \alpha;\ E(\pi) \geqslant 0;\ \text{Max congruence};\ \text{Max } E(\pi)\}$$

$$\text{(survival)} \qquad \text{(stress)} \qquad \text{(congruence)} \qquad \text{(profits)}$$

The idea is that the landed élites have a congruence goal in the T, K_T space, since the sources of growth that substitute for land (such as the diffusion of land saving technologies and infrastructures such as irrigation, that allow the opening of new land) threaten the social power deriving from their monopoly on the land. Hence the objective function includes, in descending order of dominance: a risk aversion goal, in the form of a survival constraint; the objective of avoiding negative or falling profits (stress); the congruence goal and the profit maximisation goal.

The congruence goal creates a divergence between actual and latent demand.[42] Such a gap, according to de Janvry, may explain why, for instance, yields in the Argentine cereal sector at the time of analysis were about half of US levels, although they had been equal in both countries before the 1930s.

Second, a number of factors can shift both latent and actual demand away from the socially optimum advanced technologies. In less developed countries internal, policy-related price distortions, farmers' risk aversion and higher research costs per unit of output affect latent demand by shifting the unit cost line in Figure 4.6. Furthermore a lower stock of scientific knowledge implies an innovation possibility frontier less close to the origin than in the case of developed countries.

In a later contribution, de Janvry *et al.* (1989) state that the optimal technology choice becomes conditional on the size of the farm when transaction costs are introduced within the induction model. This is because, as the ratio of hired to family workers increases, the possibility of opportunistic behaviour – hence supervision costs and the price of a unit of effective labour – increase. Furthermore, fixed costs in land transactions imply that the price of land tends to decline with farm size. Based on the Hayami and Ruttan idea that technological advances can be classified as land saving or labour saving, the production function in agriculture is represented as a separable, two-level production function, with high elasticity of substitutions between subfunctions (the typical inputs of the green revolution – improved seeds, fertilisers, insecticides and irrigation – can easily substitute for land; while machinery and equipment substitute easily for labour) and a low or negative elasticity of substitution between the land and labour indexes.

$$Q = F[X_A(E_A A, E_F F), X_L(E_L L, E_M M)]$$

where: Q = agricultural output; X_A = 'land' input index; X_L = 'labour' input index; A = arable land; L = labour; F = land saving capital (fertiliser); M = labour saving capital (machinery); and E_i = efficiency parameters.

The additional hypotheses introduced are that at farm level the price of labour is increasing in L and the price of land is decreasing in A

$$w = w(L) \quad w' > 0 \text{ and } w'' < 0$$

$$r = r(A) \quad r' < 0 \text{ and } r'' > 0$$

and that credit availability, determined by the size of the land owned and which could be used as collateral, $K(A)$, constrains farmers' total expenditure on inputs.

In this model, since farmers of different size face different factor prices, optimum factor ratios vary with farm size. If the state allocates resources to research with the objective of maximising sectoral output a larger average farm size is associated with the choice of a more labour saving technology; and inequality in the distribution of land introduces a further bias in favour of mechanisation.

Since there are a multiplicity of private optima, the demand for the public good of agricultural research becomes social-group-specific and collective action takes the form of a distributional struggle. If the effectiveness of lobbies were inversely proportional to the size of class membership, technological biases may be further distorted towards the interests of the large farmers. On the contrary, the size of the research budget (per acre of arable land) tends to increase the technological bias towards land saving technological change. Ultimately, the distribution of assets and the size of the research budget become important determinants of the direction of technical change.

Food entitlements and the green revolution

The term 'green revolution' is used to indicate the massive introduction of high-yielding varieties of grains, initially developed for production in Mexico (wheat and corn) and the Philippines (rice), in other countries of Asia, Africa and Latin America.

The experiment was originally a response to the threat of food shortages in the Indian sub-continent in the mid-1960s, and it was indeed a successful response. In India the innovation was massively adopted and by the beginning of the 1970s wheat production had more than doubled and the country became self-sufficient in wheat.

The debate about the impact of the green revolution has been extensive, reflecting questions of agricultural development policy. At issue is not the superiority of the technology as such but rather the possibility to use it as the key element of a development strategy.

During the 1950s and 1960s there was a significant intellectual consensus about the need to deal with the structural problems of agriculture and especially with the unequal distribution of land ownership. The green revolution shifted attention away, it was perceived as the success of the 'high input' model, of which the IIM represents the most accomplished articulation, focusing the attention on technology and price signals.

Indeed, by the 1980s, emphasis on the price system, and the identification of price distortions as the major obstacle to agricultural development became 'orthodoxy' and the theoretical underpinning of the action of the World Bank and the IMF (Cuffaro 1988). Such emphasis is an intellectual extension of the belief that market forces can go a long way in solving problems, development problems, or, by extension, institutional or even population problems.

An extensive body of research has analysed – with conflicting findings – (Freebairn 1995) the impact of the introduction of the modern varieties on income distribution in three distinct dimensions: among farms of different size, according to the type of ownership (tenants versus owners), among regions with different physical and institutional infrastructure, and on landless labourers.

According to a review of studies on India (Prahladachar 1983) the green revolution spread independently of farm size and tenurial status, although in some areas small farmers' adoption was constrained by capital and unequal access to credit and to infrastructure, and although large farmers disproportionately benefited from the new varieties. Regional disparities however increased, since unequal physical and institutional infrastructure resulted in different rates of adoption of the new varieties across regions. As for the use of labour, it appears that the introduction of the high-yielding varieties required higher labour input (due to greater utilisation of labour per unit of cropped area and/or higher cropping intensity) per unit of land, provided either that the traditional techniques (those that depended only on unassisted human or human and animal power, along with simple traditional implements) were retained or the shift in cultivation techniques was limited to intermediate techniques (substituting improved implements and equipment for the traditional ones but retaining the same power sources).

In a static sense it is sufficiently non-controversial that the spread of the green revolution varieties does not necessarily require the use of mechanised equipment and that the impact of such varieties on labour use depends on the level of mechanisation with which they are cultivated. However, the dynamic relationship between the introduction of the new varieties and the spread of mechanisation seems to be less clear, and the overall impact of the green revolution on labour use is therefore ambiguous.[43]

5 The response of institutions

Institutional responses and the success of the green revolution

The green revolution was based on the development of crop varieties by international centres publicly financed and their transfer to the national research systems in the developing countries.[44]

This technology development was driven by the notion that genetic improvement is the key to rapid productivity gains in agriculture – a dominant paradigm in agricultural research for much of the post-war period (Lynam and Blaikie 1994) – and the innovation consists in the use of varieties which yield higher than the indigenous varieties at high level of application of fertiliser, irrigation and biochemical programmes for disease, insect and weed control. For instance, with the new varieties, yield increases associated with the use of chemical fertiliser cease at levels of applications three times higher than in the case of traditional varieties.[45] It is indeed the type of land-augmenting innovation process described by the move from I_t to I_{t+1} in Figure 4.6b.

Historically, the green revolution is the result of a number of developments both on the side of the demand for innovations (in developing countries) and of the supply of innovations – (mainly in the developed countries) which accelerated after the Second World War (Hayami and Ruttan 1985; Hayami and Otsuka 1994).

On the demand side, many developing countries, especially in Asia, were experiencing rapid population growth and a substantial exhaustion of the land frontier, i.e. of the possibility to expand cultivation to new land. Furthermore, a general effort at industrialisation – through import-substitution policies – in the newly independent states required low food prices for industrial workers through increasing domestic food production.

On the supply side, there existed a large accumulated gap between developed countries and the developing countries. In the former the systematic application of science to agricultural production had started (relatively late compared to industry) in the 1850s and had established itself as a major source of growth in agricultural output by the 1950s. In the latter the systematic application of science to agricultural production had been limited to export crops under colonial rule. Furthermore, the relative price of fertiliser had declined as a result first of increased productivity in the fertiliser industry in developed countries, later of the growth of domestic fertiliser production in the developing countries through technology transfer. Within this framework it is understandable why the green revolution succeeded mainly in Asia, where population pressure on land, and therefore presumably the rice–fertiliser ratio, were higher (Hayami and Ruttan 1985; Hayami and Otsuka 1994).

Hayami and Ruttan have conceptualised the green revolution as the shift from point A to point B in Figure 5.1, where u_0 and u_1 represent the fertiliser response curve respectively for traditional and modern varieties and *U* is the envelope of such response curves, termed the 'metaproduction function'. If the fertiliser–rice price ratio moves from p_0 to p_1 individual farmers move from A to C if u_1 is not available or to B if u_1 is available.

However, the change in technology is obviously more complicated than the figure shows, and its preconditions are also more complex than the change of the price ratio that triggers a technological response in Figure 4.6.

The realisation of the yield potential of the modern varieties (MVs) is linked to the application of a package of inputs which includes, with fertiliser, also water control and husbandry practices such as weed control.

> Adjustments along the metaproduction function usually involve time and costs. The development of fertiliser-responsive MVs requires investments in research. Better husbandry practices must be developed and learned. Complementary investment in irrigation and drainage may be required to secure adequate control of water. It takes time to reorient the efforts of public agencies in such directions in response to price changes. It is particularly costly and time consuming to build adequate institutions and competent research staff.
>
> (Hayami and Ruttan 1985: 276)

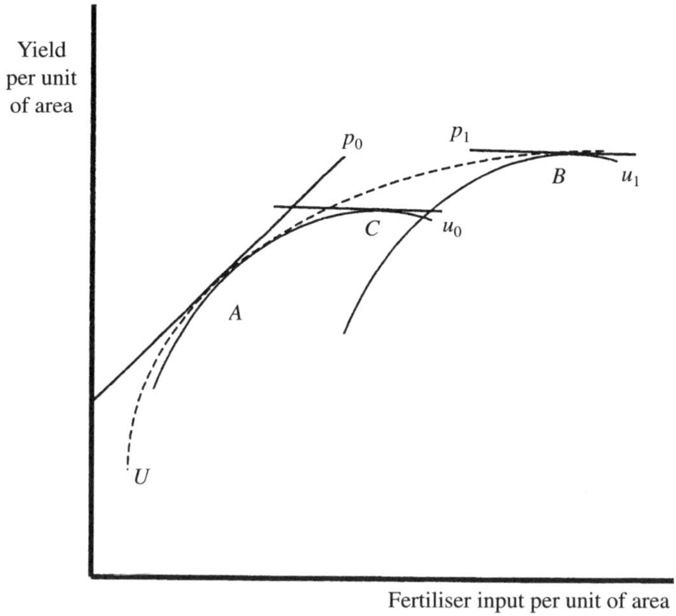

Figure 5.1 Fertiliser response curves.
Source: Hayami and Ruttan (1985)

A comparison of the response to decreases in the fertiliser–rice price ratio in Japan and in other tropical Asia countries (Hayami and Ruttan 1985) shows that in the latter, rice yields increased slowly before the mid-1960s in spite of substantial declines in the fertiliser–rice price ratio. According to the authors, the reason for such delay is that the time lag required to move along the metaproduction function tends to be extremely long in situations characterised by lack of adequate institutions and human capital to generate the flow of new techniques.

Hence, we do have two sets of related problems: one is a problem of sequencing, the other is a problem of institutional responses. Lipton observes that technological change

> cannot, economically, turn 'this Island into a garden' in any sequence one wishes. The appropriate sequences usually require both public and private investments substantially higher than are likely. . . . With some exceptions, biological, chemical, agronomic and mechanical innovations – in the absence of improved structures

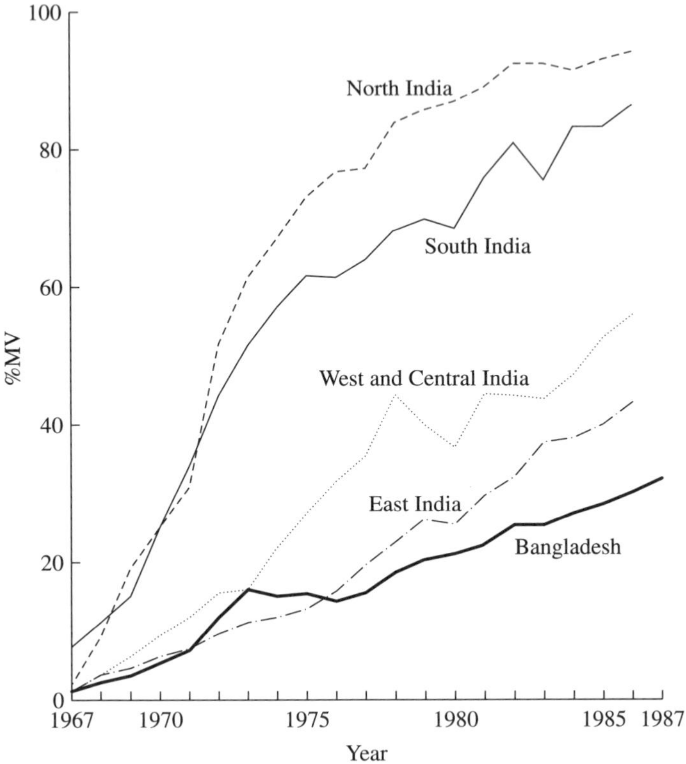

Figure 5.2 Diffusion of modern varieties (percentage of rice harvested area).
Source: Hayami and Otsuka (1994)

> for water management – have, since 1960, proved unable to support profitable farmers' decisions that permit yields to keep up with, let alone outpace, rural population growth in developing countries. But, given water control, green-revolution type responses to population growth have been rapid in many years.
> (Lipton 1997: 85)

Thus, an additional reason for the success of the green revolution in Asia is the fact that historically, high population pressure led to agricultural intensification and to the creation of an irrigation infrastructure, which, together with farmers' experience in irrigated agriculture, established an essential precondition for the diffusion of the new varieties. By the same token, the difference in rates of adoption among regions in Asia shown in Figure 5.2 would be explained by the extent of water control (Hayami and Otsuka 1994).

These differences are indeed large.

Hayami and Ruttan (1985) themselves point out that the relative pace of population growth and institutional adjustment is a critical factor. Communities first adapt to population growth by opening new lands but, as inferior lands are reached, irrigation becomes a more profitable base for agricultural growth. In turn this infrastructure development enables the diffusion of the new seed-fertiliser technology. However, there can be a substantial time lag in adjusting to the new economic conditions.

Irrigation systems often require cooperation – i.e. the concurrent effort of many – to improve water supply and control. At the same time exclusion may be difficult for technical or legislative reasons – a public-good characteristic. However, unlike the case of public goods and like many common pool resources, there is rivalry in use – i.e. use by some farmers reduces the amount of irrigation water available for others. Group action by farmers is required to solve the free-rider problems in the provision of irrigation water and for establishing rules of appropriation. This necessity arises both in local systems and in centrally managed large-scale irrigation systems that need additional, collective-choice mechanisms to meet local needs at watercourse level (Tang 1994; Ostrom and Gardner 1993). Furthermore large irrigation projects require public investments.

Such 'organisational capacity and habit grow in a rural society over time, perhaps several generations . . . the allocation of public resources is also a public process involving compromises among vested interests' (Hayami and Ruttan 1985: 312), therefore it is unlikely that government investments in irrigation will provide an immediate response to the changes in man/land ratios.

The observed delay of Southeast Asian countries in adopting new technologies, in comparison with Japan, would be the result of a difficulty of adjusting the institutions to a process of fast economic change.

> In Japan . . . the organisational capacity of local communities to mobilise communal labour for irrigation developed gradually over generations in response to gradual population pressure on available land. . . . In contrast, most parts of Southeast Asia were characterised until recently by an abundant supply of unused land,

> so that the increasing population could easily be supported by opening new land for cultivation. Although the cultivation frontiers have been closing and the man–land ratio has deteriorated rapidly in recent years, the social system moulded under the land abundant conditions does not change so rapidly. . . . Thus the cultural endowments and communal institutions have resulted in a weaker base for responding to the rapidly increasing returns to the development of irrigation infrastructure in Southeast Asia than in East Asia.
>
> (Hayami and Ruttan 1985: 322–323)

There is also a relationship between water control and the structure of asset ownership. A very unequal distribution of land, resulting in small farm size and fragmented holdings make it uneconomical for most farmers to install pumps individually; moreover inequality has implications for collective action in irrigation and drainage.

Inequality in land ownership implies that on the one hand some – the large landowners – have a strong incentive to contribute to collective action, because they can appropriate a large share of its benefits; on the other hand however, the shares of small farmers and their incentive to contribute may be tiny, encouraging free-riding. The empirical literature tends to show that, on balance, inequality reduces the probability of success of collective action in irrigation (Bardhan 1993a, b; Bardhan and Ghatak 1999).

Hayami and Ruttan (1985) for instance remark that farm producers could undertake irrigation projects using seasonally idle labour. However, with inequality, each small landowner would have little incentive to act as a leader and project leadership by large landowners is likely to be seen as an act of self-interest by the community. A similar argument has been made by Boyce (1987) in reference to the difficulty of collective action for irrigation projects in Bangladesh.

Other authors have argued that there is a relationship between the asset ownership pattern and the success of the green revolution: small peasant farming based on family labour – which is more dominant in the food crop sector in Asia, as compared to Africa and Latin America – is much better suited to the modern technology than farming based on hired labour (de Janvry 1981; Pingali *et al.* 1987). Both de Janvry (1973) and Boyce (1987) have documented, respectively in the cases of Argentina and Bangladesh, that the time lag in the induced innovation mechanism, linked to an unequal structure of asset ownership in agriculture, can be substantial, and can therefore depress the long-run growth rate of agriculture and of the standard of living.

Property rights on land

The conventional view: traditional tenure systems as dysfunctional institutions

A system of property rights on land defines the uses of land – cultivation, grazing, hunting, right to plant trees and so on – which a society or group regards as exclusive, and who has these exclusive rights. In addition, a system of property rights provides a set of rules governing inheritance and transferability. Property rights are defined and enforced through institutional arrangements that include both formal procedures (courts, police, the legal profession, land surveys, record keeping systems and titling agencies) and the social customs and attitudes concerning the legitimacy and recognition of those rights.

In many developing regions, including most countries of Africa, indigenous areas of several Latin American countries, and China the dominant type of property right system is some form of communal tenure,[46] where land rights are defined for the group and the rights of individuals or families belonging to the group rest on customs, enforcing group control over the use and disposition of land.

The conventional analysis of communal property and the ensuing policy prescriptions have been largely based on the views of the property rights school (Demsetz 1967; Alchian and Demsetz 1973), and the first argument against the communal type of rights is a resource degradation – 'tragedy of the commons' – argument, whose structure is discussed in Chapter 3.

A second argument is the claim that customary land arrangements would be a hindrance to intensification, for four reasons. In the first place the insecurity of individual land rights would reduce farmers' incentives to make long-term investments in land improvement and conservation, because such benefits might be appropriated by other persons and/or would not accrue to their own children. Data from farming households in several regions of Rwanda (Migot-Adolla *et al.* 1993) support this view, showing that parcels that could not be bequeathed were much less likely to be improved by farmers in any manner. Also, the very rules of group control might limit farmers' ability to make intensification investments.

A further question is whether customary land arrangements, by reducing the scope for land transactions and land rental, reduce the scope for efficiency-enhancing transfers. Atwood (1990) observes, for Africa, that although transactions, including fairly common sales, take place in a wide variety of land tenure systems, customary land tenure systems may introduce substantial risks and transaction costs which lower potential buyers' expected return from the land. Also, land rental exists (albeit often not recognised or sanctioned in government land laws), but insecurity on the part of landowners, who perceive substantial risks of losing their land if they rent it out, may reduce the extent of the rental market.

Moreover, a system providing every resident household with as much land as it needs for subsistence, but no more, would prevent households from producing a surplus. As Platteau (1992) observes, this last argument ultimately implies that traditional land tenure arrangements in Africa may prevent the process of gradual improvement of the quality of the land (and the productivity of livestock herds) over generations which occurred in other parts of the world, particularly in Europe and Asia.

Finally there is an access to credit argument, supported by some empirical evidence on Africa and Asia: insecure legal ownership of land, by reducing the possibility to use land as collateral, would reduce farmers' access to credit.

In a 1987 survey on sub-Saharan Africa, Feder and Noronha (1987) found that secure legal ownership has an important role in providing farmers with access to cheaper, long-term, and more extensive credit; that title may also increase the supply of all type of credits (including informal credit) and that the effects of constrained and dearer credit are significant. Feder and Feeny (1991) compared the performance of squatters on state land, who lack titles on land they farm, with that of titled farmers in Thailand:

> The results show that titled land . . . bore little risk of expropriation, provided better access to credit, and had significantly higher market value as compared with squatters' land. Titled farmers had a larger volume of investment, higher likelihood of land improvements, more intensive use of variable inputs, and higher output per unit of land.
>
> (Feder and Feeny 1991)

The reason is that the availability of land as collateral and documentation of land rights that make such collateral credible affect the willingness of creditors to make loans. In addition, formal procedures for registering liens on property rights provide important enforcement mechanisms. Thus, the same institutional arrangements that increase incentives for productive use of land also facilitate a more efficient credit market.

In rural areas of developing countries typically the formal credit market coexists with informal credit markets. In the first, formal institutions provide intermediation between depositors (or the government) and lenders at relatively low and usually government-subsidised rates of interest; in the informal market money is lent by private individuals at extremely high rates of interest.

The high interest rates prevailing in informal credit markets have received two types of explanation: in one view the village moneylender charges usurious interest rates because he can act as a monopolist; in another view (Stigler and the Chicago School) credit markets are approximately competitive and high interest rates reflect only high risks of default and high costs of information (hence the only ground for government intervention would be to redistribute income to the poor).

As Hoff and Stiglitz (1993) observe, neither theory however provides an explanation for a number of features of rural credit markets, including the fact that the formal and informal sectors coexist, despite wide differences in interest rates, and the fact that formal lenders tend to specialise in areas where farmers have land titles. More recent interpretations see the structure of rural credit markets as the consequence of problems of imperfect information and enforcement (Hoff and Stiglitz 1993): lenders have the problem of screening among potential applicants and of ensuring that loans are actually repaid. This has a number of consequences. For example, the interest rate has the dual function of rationing credit and regulating the risk composition of the lender's portfolio, since the mix of perspective projects tilts in favour of riskier projects as interests increase. Hence, while orthodox economic analysis suggests that interest rates reach a level which equates demand with supply, when imperfect information is introduced the demand for credit may actually exceed supply, while lenders choose to keep their rates low and ration the available loan funds through other means.

It appears however, that the information and enforcement problems which lenders face in rural credit markets of developing countries are mostly solved through direct screening mechanisms based either on geography and kinship (the cost of information for potential borrowers is low if such borrowers are neighbours or part of the same kinship group)[47] or on interlinkages with other markets (e.g. lenders who are also non-resident traders generally require that their clients sell all their crops to, or through, them). Furthermore collateral requirements, usufruct loans, rotating savings and credit associations, which are common features of rural credit markets in developing countries, can be interpreted as methods to limit the consequences of information asymmetries and enforcement problems. Unlike informal lenders, banks rely heavily on collateral, generally in the form of land, because of greater difficulty in the direct screening and monitoring of borrowers.

Extensive asymmetries of information and limited availability of collateral and other means of enforcement tend to generate a segmented market – because different lenders face different possibilities and costs in relation to screening and enforcement – and tend to result in monopolistic competition within the informal credit market (Hoff and Stiglitz 1993).

Although the imperfect information approach can explain several features of rural credit markets, ultimately in relation to our subject, i.e. communal property rights, the ensuing diagnosis (and policy prescription) does not significantly differ from that of the property rights school:

> The creation of a dense network of market interactions, which we would expect as development proceeds, lowers screening and enforcement costs. Legal developments such as land titling, in conjunction with the individualisation of land rights as commercialisation proceeds, allow land to be used as a collateral, and that in turn expands the scope of credit markets . . . Land titling, to the extent it increases the value of land as collateral, and the introduction of cash crops, which makes possible interlinked trade-credit contracts, will reduce lenders' costs of enforcement.
>
> (Hoff and Stiglitz 1993: 46–49)

Critique of the conventional view

The set of arguments outlined in the previous section has been the ground for policies supporting the creation of freely tradable private property rights on land: i.e. the abandonment of communal tenure systems in favour of freehold title, the subdivision of the commons on the basis of land titling programmes.[48]

The conventional view has been subject to considerable debate and revision, as research showed that it is in many ways based on an inadequate understanding of the nature and consequences of traditional tenure systems.

In the first place, since the debate about the relation between communal systems and agriculture in developing countries has emphasised the issue of insecurity of individual land rights – as leading possibly to great externalities and to a 'tragedy of the commons', or as a hindrance to intensification – one must assess the empirical evidence regarding the nature of such systems. The use rights defined in customary tenure systems often have no legal status and are typically conditioned by group rights or secondary rights. Nevertheless, the evidence clearly shows that, unlike open access, communal property implies that property rights are defined, although they are exercised (fully or partially) by a group rather than by an individual, and that membership of the group is limited. Indigenous systems usually ensure individual cultivators security of possession over a specific plot for a definite purpose and length of time, while transferability is limited in various degrees.

As Otsuka and Place (2000) correctly observe, open access, rather than a system of property rights, is an extreme case of deterioration which may occur not only with common property but also with state property: for example, because of the inability of governments to enforce property rights, state owned forests are often open access.

The distinction between security and transferability is especially important. Security refers to a farmer's use rights on a parcel of land, i.e. to the ability to use land for a certain period, or for lifetime, and for a defined purpose without disturbance. Transferability on the other hand refers to the norms that regulate the alienation of the land. When security and transferability are used as two distinct dimensions to classify systems of property rights on land, as in Figure 5.3, a system where no individual property rights are defined would occupy the origin O while the private property system would occupy point P.

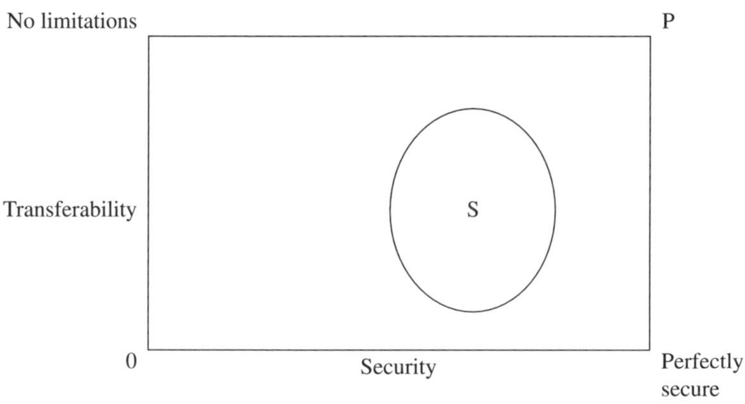

Figure 5.3 Communal property.

Source: adapted from Hoff (1993)

Most communal systems of property rights on land would belong to a space such as S, where a substantial degree of security is coupled with limited transferability.

This is approximately the area occupied by the African indigenous land right systems analysed in Migot-Adholla *et al.* (1993), who collected data on whether heads of households could exercise numerous use and transfer rights over particular parcels of land. The specific rights enumerated for each parcel included use rights (rights to grow annual crops, to grow perennial crops, to be buried, to make permanent improvements and to collect firewood, among others) and transfer rights (right to lend, rent, pledge, mortgage, give, bequeath and sell). According to their findings, use rights were generally quite secure, while the possibility to transfer such rights was in many cases limited by the requirement: (a) to obtain prior approval; (b) to limit the transfer to short duration; and (c) to make transfers only to someone within the family, lineage or tribe.

Hence, when distinction is made between individual right to use and right to transfer, it appears that indigenous systems usually ensure individual cultivators security of possession over a specific plot for a definite purpose and length of time, while transferability is limited in various degrees.

These findings and other evidence on developing countries (Feder and Noronha 1987; Deininger and Binswanger 1998) disprove the simplistic dichotomy between private property and common property in land and the description of the latter as an open-access situation.

The theme of cooperation in common property resources problems in general has already been discussed in Chapter 3, where it was shown that cooperation is possible and the tragedy of the commons outcome is not inevitable.

As for the dynamic efficiency arguments against traditional land tenure systems, Deininger and Binswanger (1998) observe that, under conditions of low population density, there is little economic rationale for land-related investments, independently of the system of tenure. In addition there is evidence that tenure-related investment disincentives are attenuated in many systems, as they recognise property rights to land improvements (e.g. trees) or compensate for improvements made upon redistribution of land. Sjaastad and Bromley (1997) observe that an African farmer may be compensated for any trees planted, buildings erected, or furrows dug when he loses an ownership dispute and that, although communal systems prohibit land transactions with outsiders, rental and often even sale within the community are normally allowed and do provide scope for efficiency-enhancing transfers. The same authors point out the lack of congruence of the main propositions of the conventional view: 'the propositions that indigenous tenure provides insufficient investment incentives, and at the same time leads to rent dissipation, are contradictory. Each may have some merit on its own, but both cannot hold at the same time. The very act of rent-dissipating capture is largely analogous to overinvestment.'

As for the access to credit argument, it could be observed that many long-term investments in land improvement and conservation, such as those required by a gradual intensification of agriculture along the lines of the Boserup model, involve an increasing use of labour with little or no external inputs, and therefore with little requirement of external financing. Furthermore, in many developing regions informal lending predominates and informal, unregistered land claims may be accepted as collateral if the lenders are close enough to the community to have low-cost information on the legitimacy of the land claim and if they can foreclose on land when a borrower defaults. Moreover, legal title does not suffice to insure the use of land as collateral if social customs make it practically impossible for banks to foreclose on land securing a defaulted loan, as in many parts of Africa (Atwood 1990).

Finally the conventional view has been criticised because of its policy prescriptions, since privatisation disenfranchises the poorest and nationalisation often does not translate into better resources management. In many African countries land titling had deleterious equity implications because of land grabbing by influential individuals during the titling process. A further example comes from Bolivia's history of arbitrary allocation of frontier lands to influential individuals: titling was highly arbitrary with the poorest groups (including the indigenous population) being least able to protect their rights (Deininger and Binswanger 1998). In the context of Africa, Atwood (1990) notes that land registration's formal legal and administrative mechanisms – boundary delineation and recording of land claims – while reducing risks and transaction costs for some, may increase risks and transaction costs for others, especially local people who rely on existing informal means to establish and safeguard their land claims.

Communal property in land as a functional institution

> The person who cleared the land first was, in absence of any more powerful claim, entitled to use it . . . Anybody who later tried to establish rights of use within the area already cleared and

> controlled had to seek permission . . .When land was abundant access to it was not difficult. It was obtained either by residence or by acquiring 'membership' in a group . . . The admission of outsiders . . . was common. The crucial element for the continued control and use of the land was to have enough people, be they relative or slaves, to work the land. Given the primitive technology . . . the abundance of land, and the practice of shifting cultivation, land had little or no economic value. It thus made no sense to develop a system of rights to a particular parcel of land that could not be protected and had no utility until its fertility was restored. Accordingly land was under group control, and individuals used particular bits of it . . . Under such system of land use, one person could cultivate crops, while, on the same land, another could have rights to trees; or land could be used by cultivators during the crop season and by herders in the off season or during fallow periods.

This long quotation from Feder and Noronha's (1987: 146–147) description of land tenure in pre-colonial sub-Saharan Africa illustrates the rationale for communal property in conditions of land abundance.

There is a set of circumstances under which communal rights are a functional institution. These circumstances include a degree of land abundance such that great investments for restoring and enhancing fertility on a given land are not required, while at the same time it pays to exclude outsiders from using it; and a size of the community such that the transaction costs of regulating use among members are not prohibitive. In addition, communal ownership allows forms of organisation that would be difficult to achieve under private property, such as the utilisation of economies of scale to break seasonal labour bottlenecks and the realisation of investments in community-level infrastructure (Deininger and Binswanger 1998).

There is also evidence (Otsuka 2000) that complex inheritance systems – such as those where land is bequeathed from men to brothers and nephews or from women to sisters, daughters and nieces – tend to evolve in order to secure rights for those who are actually involved in the productive use of the land – i.e. wife and children. For example in Ghana, under a traditional system of inheritance from man to brother or ultimately nephew, wife and children had little incentive to provide critical labour input for cocoa-tree planting. The emergence, in this setting, of a system whereby land could be donated during life to wife and children – if they had provided labour to help establish cocoa fields – is interpretable as a functional institutional evolution, where land investments are a way to assert individual rights within communal systems.

Indeed, Sjaastad and Bromley (1997) have argued that in general, within indigenous tenure systems in sub-Saharan Africa, certain types of investment in land are a legitimate way of claiming more secure rights to land and that investments may be recovered even when land is lost. If the gains of an investment under freehold tenure are termed *Gf*, the corresponding gains for a farmer within an indigenous type of tenure system *Gi* is given by *Gf* multiplied by a factor that captures the probability of recovery of the investment[49] plus a term *Gs* that captures the increased expectation of rent capture resulting from the lower probability of eviction generated by the investment, i.e. the fact that tenure security is also a result of land use decisions. On this basis, the comparison of static investment incentives under the two tenure systems and dynamic considerations lead the authors to the conclusion that indigenous tenures may not only provide investment incentives that are superior to freehold, but also a path towards tenure security more efficient than state intervention in the form of imposed titling programmes.

Otsuka (2000) has found evidence supporting the above argument. In Uganda, Ghana and Sumatra commercial tree planting has not significantly differed between areas under communal ownership and areas with stronger individual rights. This result is attributed partly to the fact that once trees are planted the land ownership system is often converted to de facto private ownership within the community. Moreover, such institutional rules have been observed in areas with a strong comparative advantage in agroforestry, leading the author to conclude that 'Communal land tenure institutions have built-in rules to ensure the intensification of land use as predicted by Boserup in areas where agroforestry has a comparative advantage'.

Population growth and the evolution of property rights on land: the neoclassical view

Recent studies on the relationship between population growth and the evolution of land rights in developing countries have mostly used the theoretical framework of the Alchian and Demsetz (1973) analysis, and specifically the notions that the changes of property rights required to 'internalise' the externalities associated with common property imply costs (transaction costs); and that property rights develop when the gains from such internalisation become larger than its costs.[50]

Research on developing countries' recent history confirms the idea that the rise of the relative price of land induces a process of evolution from group control of land to private property and, although such a rise can be caused by many factors – including the development of markets for products, the growth of communications and population growth – this last factor has been the most frequently considered. As discussed, recent research also dismisses the simplistic dichotomy between private property and common property and the description of the latter as an open-access situation, and accordingly describes the evolution of land rights as a very complex process, involving many adjustments.

Property rights on land tend to evolve in response to changing population pressure[51] following a broadly identifiable pattern. In land-abundant environments, with shifting cultivation, where labour is the limiting factor and uncultivated plots have no economic value, individual property rights are usually not defined, land tends to be under group control and membership of the group is easily acquired. As population increases and land becomes scarcer, there is a tendency to restrict the membership of the group and to transfer various rights from the group to individuals. This, in turn, provides incentives to adopt the fertility restoring techniques and provide the physical investments (e.g. terracing and tree planting) required by shorter fallow cultivation.

Another possible institutional development is the establishment of common property regimes with stricter control of natural resource extraction for forests and woodlands (Otsuka and Place 2001).

Such processes of change involve several institutions (the formal legal system, the system of cultural norms and customary law-enforcement mechanisms) and, as Feder and Feeny (1991) have pointed out, in many contemporary developing countries changes at these different levels are not necessarily congruent. A society where the formal legal system contemplates private property might lack the corresponding registration and enforcement mechanism; the transfer of land to another clan or ethnic group may be allowed by the legal system but not by cultural norms.

Empirical evidence showing a positive association between the degree of privatisation of agricultural land and population density is provided in a wide number of studies on sub-Saharan Africa (Pingali *et al.* 1987; Migot-Adholla *et al.* 1993) and Asia (Hayami and Ruttan 1985; Feder and Feeny 1991). Findings of these studies also support the notion that the degree of privatisation is positively related to other factors as well, including the level of market infrastructure, increased commercialisation and expanded possibilities for trade.

Equity, efficiency and power

The property rights view can explain a process such as the one described by Figure 5.4. This is an ideal sequence that leads a community to a structure of agriculture based on owner operated family farms (Binswanger *et al.* 1995).

The agricultural economics literature tends to show that, in general, there are no significant economies of scale in agriculture; and that there is a negative relationship between farm size and land productivity[52] (Binswanger *et al.* 1995). Hence, the process of Figure 5.4 would, as population grows, establish the incentives not only to protect the long-term productivity of the land but also to raise yields.

The proposition that a structure of agriculture based on small, owner operated, family farms leads to greater social equity *and* social efficiency of resource use in agriculture – as compared to a structure with land concentration – is supported by considerable empirical evidence on the existence of an inverse relationship between farm size and productivity – essentially physical productivity per hectare[53] and by

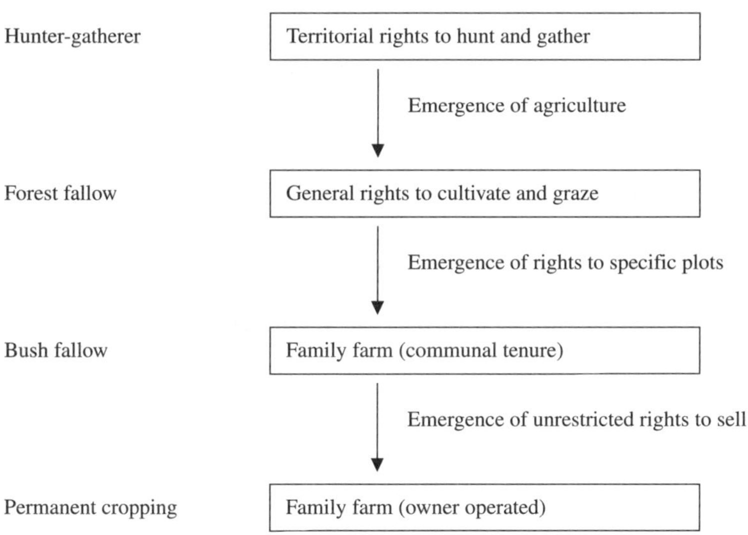

Figure 5.4 The evolution of property rights on land.

the explanations offered for such evidence in the agricultural and development economics literature.

Although an extensive review of this debate is beyond the scope of this work, its main arguments can be summarised as follows.[54] Smaller farms use the land factor more intensely, by cultivating a greater proportion of their land, using labour more intensively and therefore being able to choose more intensive types of cultivation and multiple cropping.

Lower land utilisation could partly be due to the fact that large holdings are in some cases acquired for reasons different from their use for production (e.g. as portfolio investment in high inflation countries or for social prestige), however the systematic bias is explained by the fact that farms of different sizes face different implicit prices for land and labour.

Land may be cheaper for large farmers because of better access to formal credit markets or low-priced distress sales by smallholders, or may be perceived as cheaper if it is acquired through large hereditary transfers.

On the other hand, agricultural production in small farms is more labour intensive for two reasons. One involves a hypothesis concerning the use of family – rather than wage – labour: family farms would employ labour to the point of equalisation between the market wage and the average, rather than marginal, product. With decreasing returns to labour this implies that more labour is applied to production in small farms even if they operate with the same technique (production function) as large farms. The second explanation is based on the idea that the price of labour is higher for large farms. Reasons include costs linked to 'moral hazard' problems (e.g. supervision costs) and costs that family labour faces in the rural labour market (e.g. transportation costs). Better access to formal credit markets also implies that capital is cheaper for large farmers.

In summary, factor market imperfections would lead large farmers to adopt techniques not in line with factor scarcities in the economy – i.e. more capital intensive – (and to lobby for mechanisation subsidies). Hence, the claim that small family farming achieves outcomes that are superior on efficiency, as well as equity, grounds (Ellis 1993).

Obviously however the sequence shown in Figure 5.4, and its theoretical underpinning, does not explain the great variety of the structure of asset ownership and production relations in agriculture found in history and in the contemporary world.

In contrast with the evolution shown by Figure 5.4, nearly everywhere there have been long periods of history during which a class of landlords appropriated land, imposed tributes, taxes or rent in cash, in kind or in corvèe labour to farmers on the estate and restricted their freedom by violence or the threat of violence[55] (Binswanger *et al.* 1995). The emergence of a class of landlords in medieval Europe and the manorial system discussed in Chapter 2 is an example of this.

If coercion was no longer possible or was insufficient to keep farmers on the landlord's estate a number of measures were adopted to reduce the expected utility from family farming outside the estates, including the reduction of the land available outside such estates through the appropriation of unused lands and of agricultural public goods such as roads and extension, and of credit and subsidies to agriculture.

The concentration of ownership and the predominance of large farms observed in the developing world at the end of the Second World War is interpreted as the result of such power relations and rent seeking[56] activities. Such structures have then evolved either, favourably to the peasants, towards a system of family farms or, favourably to the landlords, towards a system of large *junker* estates, with a much expanded landlord's home farm at the expense of land allocated to tenants. These estates were less efficient than family farms and their creation would be the result of the threat of land reform and of legislation protecting tenants' rights. Under the same pressure, landlords sought to expel tenants from their land and use their influence over the state to obtain protection and subsidies for mechanisation, which allowed their evolution into large-scale, mechanised commercial farms.

In their perspective the authors observe that 'If there were economies of scale in agriculture beyond those that a family could take advantage of with a given level of technology . . . it would not have been necessary to use power to aggregate large holdings or coercion and distortions to recruit workers. And in modern times it would not have been necessary to subsidise large commercial farms . . . Voluntary transactions in undistorted markets would have achieved these ends . . .' (Binswanger *et al.* 1995: 2665).

Thus, in many areas of the world, under increasing population pressure, the power struggle over the definition of property rights on land resulted in the dissipation of rents into rent seeking activities and the use of technologies not in line with relative factor scarcities.

An illustrative example of the role of power in land rights is provided by the history of Kenya's agriculture under British rule. The 1902 Crown Lands' Ordinance assigned to British settlers and British multinationals a substantial part of the best land – highlands with high rainfall – dispossessing the native traditional users of the land and creating a structure of large holdings (each settler acquired on average 2500 acres (1000 ha)) which became plantations of high value crops for export. The role of such holdings in influencing their own profitability and the profitability of smallholder farming through control of policy-making is described in El-Ghonemy (1990).

Plantation owners influenced the passage of several laws that coerced African labourers to work for them. Measures to reduce the expected utility from family farming outside the estates included the fact that native African farmers were banned from growing some high value crops. Command over policy-making translated into the appropriation of agricultural public goods: large plantations captured a disproportionate share of public expenditure in the form of infrastructural facilities and social services (although nearly 70 per cent of the tax burden was borne by native Africans) and dominated institutional support and technical agricultural services. After independence 'Africanisation', in spite of legislative efforts to expand the smallholder sector, has left the protected position of the large-farm sector substantially intact (El-Ghonemy, 1990: 161–162).

Kenya's history does not represent an exception:

> the pre-land reform concentration of large properties in Latin America, many South-East Asian and Middle East countries were a function of colonial rule. History tells us that British Viceroys, the Spanish Crown and Ottoman Sultans granted large estates to holders of certain offices and influential families on whose support the colonial rulers were dependent . . . many of these . . . holders of land dispossessed the powerless small-holders and exercised land-grabbing. At a later stage, large holdings of both origins were converted to freehold private ownership.
>
> (El-Ghonemy 1990: 85)

Christodoulou (1990: 44–45) remarks:

> One principle generally applied . . . was that 'vacant' or 'unclaimed' lands became 'Crown' or public lands, even though in customary law they belonged to lineage groups or to communities as a whole . . . Plantations first based on slavery and then on indentured labour have left a mark on the composition of the population as well as on the agrarian structure in many Latin American, Caribbean, Pacific and Asian countries . . . [In] independent

> ex-colonial territories . . . many succeeded to the lands of departed colonists or settlers . . . This early start by a small minority of élites from colonial territories is one of the key features in the development of the ex-colonial world.

Hence, the evolution of property rights on land – in the sense of the pattern of private land ownership established – has been heavily influenced by other institutions – the state and/or the colonial power – and by the power relations within such institutions. Once a heavily concentrated structure of land holdings is established it tends to set path dependency, because of the influence large landowners exert in the policy arena.

Land concentration has implications for the ability to acquire food because it tends to be associated with rural poverty and, to a lesser extent landlessness (El-Ghonemy 1990: 171–175); given the land endowment, population growth is more likely to result in landlessness where there are strong initial inequalities in land ownership; in turn landlessness could be reduced via land redistribution.

One possible explanation for the association between land concentration and poverty is that the dominance of large estates, because of their labour saving bias, results in lower demand, greater supply and depressed earnings for wage labour. Also, the landless are more likely to fall into indigence than micro-farmers in most economic environments in a typical year and there is evidence that they suffer worst in famines, most likely because even a little land increases security against fluctuations. Furthermore, the landless may have difficulties in getting employment and earning a wage. Dasgupta and Ray (1986) show that – since a person's food intake affects his productivity – those who cannot count on land to sustain their food consumption levels, and must therefore rely solely on their wage income, are 'expensive' workers relative to persons with some land assets. The landless therefore will not undercut the employed and will remain malnourished. Finally, with an equal distribution of land holdings, as population grows the growing rent would be retained by farmers instead of being paid by tenants to landlords (Lipton 1985, 1995).

The implications of the Dasgupta and Ray model are worth further discussion. Let's suppose that people are paid on the basis of the task completed, e.g. 10 dollars per acre harvested and that there is a relationship between the income earned, the nutrition level and their work capacity (e.g. the number of acres one individual can harvest in one day). In Figure 5.5 individuals are arranged in increasing order of land

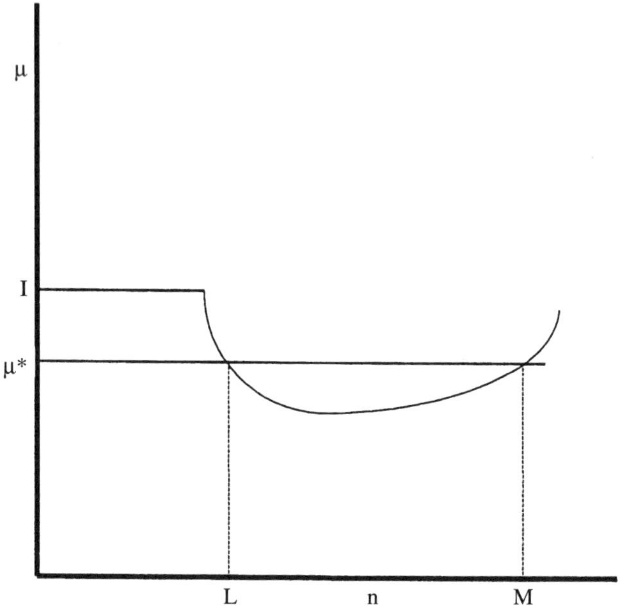

Figure 5.5 Land distribution and unemployment.

Source: adapted from Dasgupta and Ray (1986)

ownership (and therefore land income). The *y*-axis represents the minimum rate for which individuals are able and willing to work (the minimum price of a unit of labour power). Individuals with no non-labour income – the landless – cannot carry out sustained work below a minimum rate, under which nutrition is inadequate. Moving to the right along the *x*-axis, individuals with some land income can afford to work at lower rates. As non-labour income becomes high, the willingness to work becomes a constraint, i.e. individuals may have high reservation wages.

In a land poor economy, where the marginal product of labour power and therefore the competitive equilibrium level of μ may be below the level μ^* all individuals to the left of L and to the right of M are unemployed, the former because their entire food intake must be wage-based and the going market rate does not allow them to achieve adequate nutrition – they are involuntarily unemployed; the latter because at that rate they are not willing to sacrifice their leisure. Furthermore, in this setting, land redistribution would increase labour supply in the economy and output.

Hence, this model of efficiency–wage relations, based on nutritional grounds, has two implications: one is that at equilibrium there exists involuntary unemployment – a conclusion common to models that posit a relationship between the real wage paid by the employer and labour productivity (or costs), the other is that the initial distribution of assets, in this case land ownership, has implications for efficiency – i.e. the separation of efficiency and equity is not valid in this case.

Given both the observed unemployment rates in the developing countries and the large levels and apparent impact of inequality of asset ownership, these departures from the tenets of standard neoclassical wisdom – that equilibrium is characterised by market clearing, because when supply exceeds demand wages will fall; and that the problems of economic efficiency and distribution can always be separated – bring economy theory closer to reality.

Finally there is a pattern of inequality that is not class related but rather gender related. A very stimulating strand of literature has in recent years studied the intra-household patterns of resource allocation in the developing countries and their impact both within the family and on the economy at large.[57] This literature is not examined here except to add one observation to our discussion of inequality. In many rural areas of the developing countries the allocation of resources – land but also resources used for other basic needs, such as health and education – within households shows a systematic anti-female bias (Agarwal 1994, 1997). On the other hand it is acknowledged in research (Pinstrup-Andersen *et al.* 1999) that – given the level of per capita food availability – child malnutrition is inversely correlated with women's education and their status within the family. Hence, it appears that the pattern of property rights definition in many societies may undermine the positive links between food production and the nutritional status of the weak components of the household.

6 Adjustment failures

Cases of failure

While the review of past experience and of theories carried so far basically confirms the essential argument of revisionists – that population growth triggers adjustments in agriculture, both in terms of intensification and of definition of property rights – it also reveals that adaptation processes are complex and difficulties are more likely to arise when population growth rates are very fast and/or densities are already high. The 'right' institutional adjustment required by population growth may not emerge at all or not emerge on time. In agriculture, the broad historical validity of the relationships postulated by Boserup and the possibility of the adjustments foreseen by the IIM do not mean that under all conditions one may expect a smooth transition from low to high densities with sustainable intensification.

It must be remarked that Boserup herself recognised the possibility of failures in her 1965 essay, where she states:

> If it is true . . . that certain types of technical change will occur only when a certain density of population has been reached, it of course does not follow, conversely, that this technical change will occur whenever the demographic prerequisite is present. It has no doubt happened in many cases that a population, faced with a critically increased density, was without knowledge of any types of fertilisation techniques. They might then shorten the period of fallow without any other changes in methods. This constellation would typically lead to a decline in crop yields and sometimes to an exhaustion of land resources. The population would then have to face the choice between starvation and migration.
>
> (Boserup 1965: 41)

Also, in Chapter 4 the reverse linkages, running from population growth to agriculture were briefly mentioned. We do know that the increase in income and rural modernisation eventually brings total fertility rates down, but there is also evidence that this can be a long process.

> This is mainly because technical progress in the Hayami–Ruttan–Boserup style is, and ought to be, labour using – which raises the returns to child and adolescent labour well before it creates new work chances that raise the opportunity cost of child rearing for potential mothers and thus bring fertility down. The long delay also arises partly because, for the very poor, extra income at first brings significantly improved maternal nutrition and health care, and hence higher fertility (and fewer stillbirths) . . . Only after income increases have pushed poor mothers over the 'hump' of positive income effects . . . on fertility do the negative effects of extra earning prospects for women and their educated children – and hence the negative effects of 'development' on the demand for children – begin to predominate.
>
> (Lipton 1997)

Once again, the length of the process is not irrelevant, as it affects population growth and the standard of living and may change the way people affect their environmental resource base.

One effect of high population pressure is to provide the manpower required for complex land management and, when such systems are created, out-migration may provoke serious degradation. In the context of Nepal, elaborate terracing, requiring high labour input, has probably diminished erosion; on the other hand, deforestation and the creation of larger areas of grazing has probably increased it (Blaikie and Brookfield 1987). However, many case studies suggest that there is a link between population pressure and land degradation depending on a set of conditions in terms of the speed of population growth, poverty, security of land rights, initial soil conditions, and climatic factors.

Evidence from several Sahelian countries (Speirs and Olsen 1992) shows that a rapid process of increase of labour–land ratios in land abundant environments induces a deep transformation from separate crop and transhumant livestock production to mixed farming. As land is increasingly devoted to crops, farmers are led to keep their own livestock, since manuring of fields by transhumant herds diminishes, while decreasing fallowing increases, the demand for manure for fertilisation. At the same time, pastoralists have an incentive to settle and become farmers, as pasture areas shrink and their dependence on crop residue increases. However, in spite of such evolution, cropland fertilisation is not at levels sufficient to maintain soil organic matter levels, while use of inorganic fertiliser is constrained by lack of capital and by farmers' risk-averse strategies, in the context of unreliable climate. As a consequence, soil fertility is declining. In addition, the restriction of the migration areas of pastoral tribes results in overgrazing of pastureland.

Other sources (Lipton 1990; World Bank 1991) also suggest that a decline in long-term land productivity has already occurred in many parts of Africa for a number of population related reasons: rapid population growth has resulted in the reduction or abandonment of fallow periods, not counterbalanced by the use of fertility restoring techniques; the restriction of transhumant herders' mobility (as more land is converted into permanent farm land) has had negative implications for the environmental integrity of pasture lands; the increased population pressure on land reduced the ability to cope with droughts through long fallow and pastoralist mobility.

In the case of densely populated Nepal, in some areas, maintenance of soil fertility (and yields) – based on manuring from cattle that are fed forest products – is threatened by the depletion of forest resources, which in turn is due to the increasing use of these resources (for fertilisation and fuel) by a growing population (Blaikie and Brookfield 1987).

Some case studies in Africa also point out the existence of threshold density levels around which the probability of a cycle of land degradation is high. The study of a region of Tanzania (Meertens *et al.* 1996) and evidence quoted therein suggest that there is a range of population density around which agricultural labour productivity declines sharply. Communities respond first through migration, then through increased labour inputs, that is through intensification along a given production function, which in turn tends to result in soil depletion

> At these thresholds population levels (in the range of 30–60 inhabitants per km^2) the common smallholder farming systems in the vast, non industrialised areas of Africa become 'squeezed': the fields are increasingly being cropped permanently, yet neither the infrastructure, nor the local demand for farm products is sufficient to generate an economically justified intensification through the increased use of external inputs. Negative nutrient balances

> and progressive soil mining indicative of production systems that will be non-sustainable in the medium to long term therefore appear an unavoidable consequence.
>
> (Meertens *et al.* 1996: 213)

In the case study area, land regeneration and intensification based on external commercial inputs became feasible only when rural population densities reached about 100 inhabitants per km^2, and initially, in the vicinity of urban centres. The issue of threshold densities has interesting implications and will be taken up again in Chapter 7.

There may be several reasons for failure. For example lack of congruence between population densities and land potential has been evidenced in research on several sub-Saharan Africa countries (Lele and Stone 1989; Matlon 1987) and is explained by constraining factors such as colonial patterns of investment and infrastructure, barriers to migration linked to tribal history,[58] or diseases such as onchocerciasis in many of the major river valleys.

However the crucial factors for the process of sustainable intensification under the pressure of population growth are poverty and the responses of the research system and of property rights on land to changes in the underlying economic conditions. Next, the discussion focuses on the possibilities of failure linked to each of these factors.

Research

Formal research and extension systems hardly have a role in Boserup's analysis, while Hayami and Ruttan explicitly claim that such systems will (eventually) provide the right responses to changing factors' scarcity. However, national research systems may encounter significant difficulties in responding to rapidly changing underlying economic conditions, in the context of difficult physical and economic environments. For example, it is widely held that the response of research institutions to rapidly growing population in most of Africa has been inadequate (Pickering 1987; Vallaeys *et al.* 1987).

The reasons for this may lie in four different sets of factors affecting the performance of the national agricultural research systems. The first is the general low stage of development of institutions – including the nation-state – at independence and the fact that independence is a relatively recent conquest. This has far reaching consequences for 'technology'. The second is the low initial conditions in terms of scientific and educational base. The third is the nature of the African environment and of its pattern of agriculture. Fourth is the fact that the scientific paradigm that characterised the green revolution was far better suited to Asian, rather than African, conditions.

The low stage of institutional development of African states as colonial powers withdrew from the continent and the fact that independence is a relatively recent fact[59] have several implications for the national research systems (NARSs).

In the first place they have often operated in conditions of political turmoil: coups, civil wars and political unrest caused a substantial loss of agricultural scientists and teachers through brain drain in several countries, including Ghana, Ethiopia, Uganda, Somalia, Tanzania and Zambia; second, the African NARSs underwent a rapid process of destruction (of colonial research institutions) and reconstruction (of national research institutions) (Eicher 1990). Third, the fact of operating in conditions of political and/or macroeconomic instability has resulted in a relatively erratic pattern of financing.

Indeed, the history of agricultural R&D expenditures across the continent (Pardey *et al.* 1997) reveals a substantial volatility linked to macroeconomic instability. As Table 6.1 shows, real agricultural research expenditure grew at almost 7 per cent per year during the 1960s, at a much lower 2.6 per cent during the 1970s and ceased to grow in the 1980s and early 1990s as a consequence of severe economic crises and subsequent structural adjustment across the continent. But what is more disturbing is the substantial degree of cross-country variation and of volatility at country level that emerges from the data in Table 6.1.

The 1961–1991 growth rates ranged from above 10 per cent per annum for Botswana and Ethiopia, much lower values in other countries and a negative −2.4 for Madagascar. In the case of Nigeria – which is by far the largest African country in terms of population – real R&D research expenditures grew rapidly in the 1960s and 1970s, largely financed by revenues from a booming oil sector, and declined during the 1980s to a level lower than the 1971 figure.

Across Africa stagnant levels of real expenditure during the 1980s were accompanied by a continuous rise in the number of researchers. This resulted in salary crunch, widespread absenteeism in many research centres and a rapid rate of turnover of senior scientific staff (Pardey *et al.* 1997). As for the sources of funding, loans and grants from international donors accounted for around 34 per cent of total research expenditures in Africa (excluding South Africa) during the 1980s, a share that increased to about 43 per cent for a group of twenty-two countries (excluding South Africa) in 1991. These trends lead the authors to conclude that 'Many of the developments of the

Table 6.1 Africa: growth rates of real agricultural research expenditure by country (per cent p.a.)

	Annual growth rate			
Country	*1961–1971*	*1971–1981*	*1981–1991*	*1961–1991*
Botswana	30.3	13.8	−3.8	13.2
Burkina Faso	7.9	9.3	9.5	8.1
Côte d'Ivoire	5.5	1.1	0.1	1.8
Ethiopia	19.4	7.7	10.6	10.4
Ghana	4.8	−3.2	14.4	2.1
Kenya	8.4	1.7	4.0	4.4
Lesotho	20.6	6.6	−1.8	8.1
Madagascar	4.7	−7.4	3.0	−2.4
Malawi	9.9	2.4	2.4	4.0
Mauritius	9.1	1.8	1.3	4.0
Niger	8.2	12.6	3.9	6.7
Nigeria	6.4	7.1	−9.1	1.9
Rwanda	5.8	6.7	11.4	5.7
Senegal	2.9	4.7	−4.3	2.7
South Africa	6.0	−0.6	1.8	2.0
Sudan	9.9	0.5	−5.5	1.5
Swaziland	8.4	−1.2	−2.4	6.6
Zambia	14.3	4.0	0	5.3
Zimbabwe	6.3	1.1	4.2	3.6
Total	6.8	2.6	0.1	2.9

Source: Pardey *et al.* (1997)

past decade in personnel, expenditures and sources of support for public sector R&D in Africa are clearly not sustainable . . .' (Pardey *et al.* 1997: 421).

Rukuni *et al.* (1997) recapitulate the simultaneous transitions that African research managers had to tackle after independence: managerial transition from colonial to local research administrators; scientific transition from expatriate to indigenous scientists; financial transition from dependence on financial support from colonial governments and large-scale farms to mobilising support from government and donors; political transition from commercial farms to smallholders in dual agrarian societies; transition from public to private research and new forms of public/private research partnership. In addition to this, in more recent times 'Because of the failure to resolve these complex transitions, many donors are withdrawing their support and urging the managers of NARSs to pursue privatisation of research, NGO initiatives, contract research, regional research, user fees and research endowments' (Rukuni *et al.* 1997: 4–5). These latter developments pose yet another set of complex institutional issues.

The second set of conditions relates to the initial scientific and educational foundations. In discussing agricultural research in Africa during the colonial period (between 1930 and 1959) Eicher (1990) reports that a skeletal agricultural research infrastructure was established in most countries of the continent during the first two or three decades of the twentieth century. Usually, a few scientists worked on specific crops in a loose organisation more modest than present-day NARSs, while some regional research institutions dealt with export commodities for European markets. Such research infrastructure is credited by the author with some important successes and many failures, such as the inability to achieve breakthroughs in sorghum and millet after six decades of research, underinvestment in training African scientists and the pursuit of inappropriate research strategies in many land abundant countries (Eicher 1990: 127).

Pardey *et al.* (1997) point out that, at independence, African countries inherited agricultural research structures that operated as a part of a regional system, therefore many smaller countries were effectively excluded from the former network of research services, while others inherited highly specialised research agencies that did not necessarily address local production problems. The same authors observe that there was lack of congruence across countries regarding the research capacity and that research was largely oriented to meeting the demands of export agriculture and paid little attention to the production constraints faced by subsistence farming.

The third set of issues is related to the difficulty and variety of the natural environment in Africa. The physical environment for agriculture in Africa is extremely diversified and generally very difficult, especially in semi-arid areas. This is a challenge on its own but especially when considered in conjunction with the nature of the international technology transfer that took place at the onset of the nationalised NARSs.

Lynam and Blaikie (1994) remark that, in Africa, independence and the establishment of NARSs coincided with the green revolution in Asia and the founding of the Consultative Group of International Agricultural Research (CGIAR). This means that the onset of NARSs was heavily conditioned by a pattern of technology transfer based on the idea that genetic improvement is the key to rapid productivity gains in agriculture and that research had to be based on a breeding-first strategy. Hence on the one hand the defining characteristic of the African food sector is its diversity – a range of caloric staples are grown across the continent, within countries and even within individual farms, exploiting a diversity in soil, climate and terrain (from continental to farm scale) also to buffer production systems against temporal variability. On the other hand, modern plant science and well-entrenched research methods are based on the premises of the Asian model – i.e. dependence on a single crop – and tend to reinforce it.

For instance, in an analysis of the west African semi-arid tropics, Matlon (1987), in trying to explain the lack of success in introducing more productive varieties of millet and sorghum cites the difficulty of adapting Asian varieties to African conditions (due to the location specific nature of agricultural technologies) and the lack of congruence between research priorities (traditionally biased towards high yielding varieties under high-input management) and the physical and economic constraints faced by farmers.

In general, the green revolution technology, although it has had a localised success – mostly the adoption of maize varieties when maize is used as a cash crop – could not lead to the Asian type of growth and there is a problem of adaptation of established research methods to African conditions.

Finally macroeconomic and sectoral policies have been blamed for introducing a significant bias against agriculture, and therefore reducing the incentives to undertake investments – in Africa and elsewhere in the developing countries. This issue has been raised in a vast agricultural economics literature during the 1980s, including numerous World Bank studies on sub-Saharan Africa (Cuffaro 1988; Cleaver 1993).[60]

On the whole, the performance of agricultural research in Africa seems to be a further illustration of a 'timing' problem: Africa has experienced an extremely rapid process of transition from land abundance to increased population pressure on land, in the context of difficult natural environments. Society's response in terms of research could not be fast enough. Even if adaptation does occur when underlying economic conditions change, the starting conditions and the pace at which such changes occur is critical. It may thus happen that, although communities try to adapt to changing underlying economic conditions, the rapidity of change coupled with various constraints leads to a decline in long-term land productivity.

Property rights

The theory and empirical evidence examined in Chapter 3 suggest both that balanced management of common property is possible, and that property rights evolve in response to population growth. However, in specific circumstances, and especially under the pressure of fast population growth and/or when population densities are very high, the process may not go from balanced management of communal property to a complete definition of individual property rights, but rather may result in a breakdown of traditional systems into de facto open access, with the associated environmental degradation.

Theory predicts that as population pressure on land increases there is a tendency to establish private land rights, which, in turn, create incentives for the maintenance of long-term productivity. However, even if one adhered to a functionalist view by which institutions respond or new institutional arrangements emerge simply because this is called for by (and is better suited to) the new underlying economic conditions – a view that, as discussed, is subject to a wide variety of criticisms – it cannot be denied that the pace at which such underlying conditions change, and/or specific circumstances, may hamper adaptation.

One source of uncertainty is the very dynamism of property rights systems. Indeed in many parts of the developing world these systems are in a state of transition that creates many conflicts. Population pressure, ambiguous or badly administered land laws, or changing rural authority structures and institutions may give rise to controversial land claims[61] reducing the assurances formerly provided by indigenous custom. There are often conflicts between national systems of land rights, de facto rights of occupancy exercised by squatters, and customary land rights established by ethnic communities. A society in which the formal legal system contemplates private property might lack the corresponding registration and enforcement mechanism – in developing countries, official land records for rural areas are typically incomplete or do not exist at all – or, the transfer of land to another clan or ethnic group may be allowed by the legal system but not by cultural norms. Examples are provided in Atwood (1990) of African countries where formal titles are not enforceable and land continues to be allocated along lineage in spite of the existence of formal titles to land because such titles lack legitimacy. Migot Adholla (1993) observes for Africa that the indigenous tenure systems are frequently absent or very weak in land settlement areas, or following periods of major economic or political upheaval, particularly if traditional lines of authority have been severed, and that in areas where large numbers of migrants or strangers have settled and established rival claims to land owned by indigenous groups the incidence of land disputes is high.

As already discussed, traditional land tenure systems are not open-access situations, but rather common property regimes where use is regulated by a set of norms that may prevent the 'tragedy of the commons'. Population growth, by changing relative prices, should induce, and historically has in fact typically induced, a more precise definition of property rights on land. Common property, however, may collapse into open access in several circumstances, most of which include population growth as the leading factor.

Bromley and Cernea (1989) suggest one possible mechanism, the simplest, whereby egalitarian communities may experience population growth coupled with 'the unwillingness of the group to evict redundant individuals when that eviction will almost certainly relegate the evicted to starvation'. Pingali (1990) reports several case studies which show how uncertain land rights in societies experiencing rapid population growth, by reducing the incentives for maintaining soil fertility, result in a soil mining process whose ultimate outcome is a low yield equilibrium with outmigration of population.

In a simulation study of the Sahelian drought in the 1970s,[62] the existence of a crucial common property resource, grazing land, coupled with a very uneven distribution of rainfall, generate a mechanism where rapid population growth is associated with environmental crash. Human and animal populations build up during periods of good rainfall; overgrazing and land deterioration ensue: in this context a drought, coupled with people's effort to maintain themselves and their herds, could cause an economic and environmental collapse.

Finally, rural–rural migration may be the source of much conflict since, with a rapidly growing number of 'strangers' to the community, traditional norms may fail to provide adequate protection to land that is not already in the farming cycle, and hence, not under individual or family management (World Bank 1991). Bardhan (1993a, b) observes that migration and mobility – which are often linked to rapid population growth – tend to work against some of the conditions for cooperation found in the repeated game literature discussed in Chapter 3: they reduce the effectiveness of social norms and the validity of the common-knowledge assumption, and render prolonged repetition of the game more uncertain, raising incentives for short-run opportunism.

Poverty

There is a strong relationship between poverty and possible failures of adjustment to population growth. Lipton (1990), addressing the link between necessity and invention, which is at the basis of the Boserup model of induced innovation, points out that necessity must not be hunger alone, but rather hunger backed by effective demand, and that this may not be the case in very poor, intensive-agriculture regions, with severe income inequity. The examples he quotes are some of the most backward areas of eastern India and southern Bangladesh, where further intensification may be very costly and the growth of effective demand too slow to provide incentives for Boserupian technological shifts.

In general, sustainable output may be lower than the maximum attainable in the short run, while farmers or pastoralists may not be able to postpone production if they are merely achieving subsistence. Poverty also reduces the options available to farmers in terms of substitution of man-made inputs for natural resources and in terms of investments in resources maintenance.

While it is true that the poor tend to be both the victims and agents of environmental damage (World Bank 1992) this does not imply that the poor cause most environmental damage. Reardon and Vosti (1997) illustrate this point through the examples of the tropical forest in the Brazilian Amazon, where millions of hectares were devastated by large landowners using sophisticated technology, and of the Sahel, where richer households owning much livestock bear the greatest responsibility for overgrazing.

The same authors point out the fact that poverty is relevant to the extent that welfare-poor households may also be 'conservation-investment' poor, i.e. they may lack the ability to 'make minimum investments in resource improvements to maintain or enhance the quantity and quality of the resource base, to forestall or reverse resource degradation' (Reardon and Vosti 1997). A number of area-specific case studies show that such inability may relate to different types of poverty. For example, in the Sahel, households have access to land, but land quality is low. Pressure on the fragile margins could be reduced if cropping were intensified but poor households lack cash and credit to purchase non-labour variable inputs and poor communities lack complementary infrastructure (such as roads and depots) to obtain these inputs. Also, protecting fields against flash floods and runoff through bund construction requires expensive complementary materials.

Another example quoted is the Rwanda highlands, where population pressure on land is extremely high and the distribution of non-farm income very skewed. Households with little land, animals and off-farm capital are forced to pursue labour-led intensification with little use of fertiliser or mulch and manure. This leads to very rapidly decreasing returns and eventually to soil exhaustion, generating a vicious circle of poverty- and labour-led intensification, degradation and more poverty. Finally, in the case of the tropical rain forests of Brazil, where poor communities practise slash and burn agriculture (i.e. forest is converted into farmland by burning trees to generate nutrients to enhance soil), the poor have access to resources – forest cover and biodiversity – which are globally scarce, but the markets that could translate such global scarcity into income streams for farmers are missing (Reardon and Vosti 1997: 59–60).

Resource degradation processes, tend to result under complex circumstances including, aside from population growth, land concentration and inappropriate public policies. Increasing population pressure tends to induce privatisation of land, but this is likely to involve a certain degree of land concentration in the hands of the more powerful groups in the community. Where resources are commonly owned, disruptive pressures on the commons are likely to occur within processes of social differentiation whereby some groups appropriate resources while others become increasingly reliant on a shrinking resource base. Public policies may play a role in this process, as the state attempts to establish formal management systems to replace customary ones – often with very poor results – and as former common or open-access resources are transformed into state property.[63] Thus, a process of concentration of the more productive resources in the hands of traditionally powerful groups within communities – or into those of outside interests with strong links to the administration – may result in the disenfranchised poor placing increasing pressure on the residual commons, in the context of conflicting formal and informal management rules.

Moorehead (1989) describes the breakdown of a system of management of common-property resources that had evolved over hundreds of years in the inland Niger Delta of Mali, as the result of a combination of the factors quoted above and of a prolonged drought that diminished the resource base of the area. A process where inequality in access to land (and the investment pattern of large landowners) has critically reinforced the negative effects of population pressure on the environment, is outlined in a study of a region of Honduras (De Walt *et al.* 1993) characterised by rapidly growing population and highly unequal distribution of land.[64]

Furthermore, when land appropriation results in concentration of land in the hands of few, rental and share arrangements are likely to emerge. If such arrangements imply short-term use rights, they reduce the incentive to make long-term investments that prevent land degradation (Clay *et al.* 1994).

An extreme example of Malthusian checks is given by André and Platteau (1998) who report findings from a densely populated area of Rwanda, where competition for land has resulted in increasingly unequal land distribution and pervasive incidence of land disputes. The authors establish a connection between these processes and the 1994 civil war.

A self-reinforcing mechanism between environmental degradation, poverty and high fertility rates has been hypothesised by Dasgupta and it is partly confirmed by empirical evidence on some developing regions (World Bank 1991; Dasgupta 1993, 2000). Poverty induces high fertility rates in developing countries because, in the absence of public health services and old age security, children are insurance goods. The scarcity of capital and of environmental resources (mainly water and fuelwood), with associated low productivity of labour, may provide an additional motivation for high fertility rates, since children are needed as workers and thus become also producer goods.[65] The resulting population growth damages the environmental resource base, to the extent that this base consists of unprotected common property; this, in turn, is likely to provide further private incentives for large families. Thus, whilst Boserup's thesis implies that households confer an external benefit on the community when they reproduce, this may not be the case when population size is large in relation to the local environmental resource base. In these circumstances, even though a high demand for children may be rational for individual households, population externalities are likely to be negative (Dasgupta 1993: 358–359).

7 Environmental implications

Land degradation and the Boserup model

The above discussion indicates that the possibility of declines of long-term land productivity or of cycles of land degradation must be included in a model of the relation between population growth and agricultural intensification.

The existence of threshold density levels and the process outlined by Meertens *et al.* (1996) can be represented as in Figure 7.1, where yield on a fixed amount of land is represented as a function of labour and of the stock of soil fertility. Fs_1 and Fs_2 are production functions associated respectively with stock of soil fertility s_1 and stock of soil fertility s_2 ($s_1 > s_2$). Increasing use of labour over time along Fs_1 results in increasing yields, albeit with diminishing returns, and a constant level of fertility, as labour is also used to restore fertility (fertility, unlike biological resources, has no natural growth rate but can be renewed through labour and other inputs). However, beyond a point, increasing effort with an unchanged technology may also lead to a drop of fertility, and the relevant production function becomes Fs_2.

Thus, around a certain range of density, the production function may collapse into a function associated with a lower stock of fertility. Such collapse would be possible not only in the context of open access or uncertain property rights, but also with well defined property rights on land: initial soil conditions may be poor, production may not be postponed if farmers are merely achieving subsistence, poverty may constrain the use of new inputs and the impact of agricultural practices on soil quality may not be observed and understood in time.

As density increases, a new technology represented by $F's_2$ may be introduced, but the outcome of such introduction depends on whether a depletion process has already taken place and to what extent. In Figure 7.1 the new technology is depicted as bringing on a path that

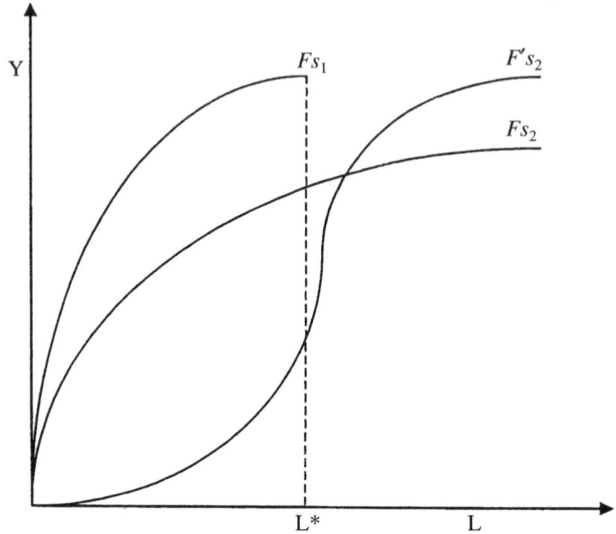

Figure 7.1 Soil fertility stock and agricultural intensification.

simply enables to keep yields constant, while in the standard version of the Boserup model (as presented in Chapter 4), there is a family of per hectare production functions, with new technologies becoming superior beyond given densities, and allowing a continuous increase of yields.

$$Y = F(L, S)$$

$$S = S_1 \text{ when } L < L^*$$

$$S = S_2 \text{ when } L > L^*$$

$$S_1 > S_2$$

where:

Y = yield on a fixed amount of land

L = labour

S = size of the resource stock (soil conditions)

$F's_2$ can be thought of as a new fertilisation technique, which includes chemical fertiliser as a supplement to other, labour intensive, means of fertilisation. The introduction of commercial inputs is constrained by the availability of capital. The Meertens *et al.* (1996) study suggests that, beyond a threshold level, high density of population and infrastructure, resulting in a larger market for farm product, will have the effect of releasing the capital constraint. This suggestion is confirmed by a finding of an extensive review of the rural non-farm labour market in Africa (Reardon 1997), showing that the more dense are infrastructure and population, the greater are farm households' earnings from the rural non-farm sector.

A more complete representation of the soil depletion stage of the process is given in Figures 7.2 and 7.3.

$$C = wL$$

$$R = PrY$$

where:

C = cost of harvesting

w = unit cost of labour ($w_1 > w_2$)

L = labour

R = total revenue from harvest

Pr = price of harvest (assumed constant)

Assuming that price is constant, the revenue-labour functions of Figure 7.2 have the same shape as the yield-labour functions of Figure 7.1. If farmers maximise profits, a private property (or group control) equilibrium (point P_1) occurs where marginal revenue and marginal costs are equal (maximum $R-C$). Open access equilibrium occurs where the total cost line ($C_1 = w_1L$) intersects (equals) total revenue (point O_1).

If population growth implies that the labour supply increases faster than demand, it will also determine lower wages and a lower opportunity cost of farm family labour. In terms of Figure 7.2 the total cost

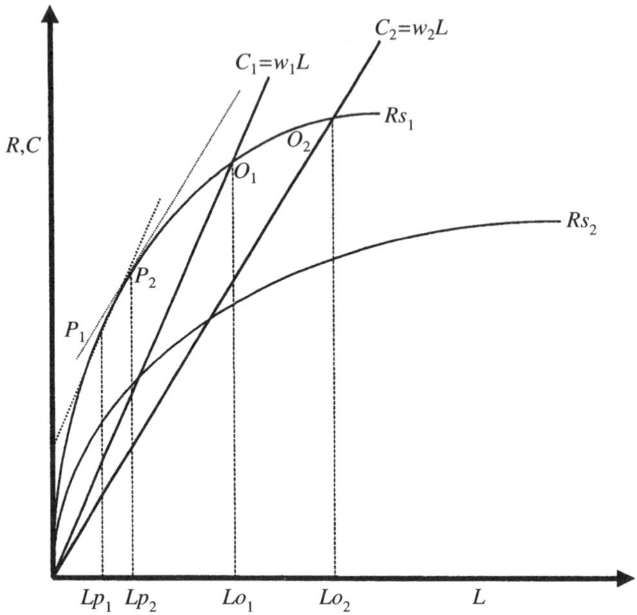

Figure 7.2 Private and open access equilibrium with different cost functions.

line C rotates towards the x-axis, resulting in a new private (P_2) or open access (O_2) equilibrium.

In Figure 7.2 points O_1 or O_2 would be also private equilibria if a family (or community) paid members their average (rather than marginal) product. In this case members would remain on farm until average product is above the prevailing market salary.

Alternatively one might refer to a family or a community where labour is considered fixed (Figure 7.3), because of social norms and lack of alternative employment, and equilibrium results from the maximisation of total revenue given L. In this case a larger size of the family or the community (from L_1 to L_3) may result in a move from A to F.

In all cases it should be noted that at lower per capita product equilibria there is also likely to be less possibility of purchasing external inputs and/or postponing production.

Thus, a larger population may imply movements along a given function that bring it close to an area where soil depletion may occur, or may result in production along a function associated with a lower stock of fertility.

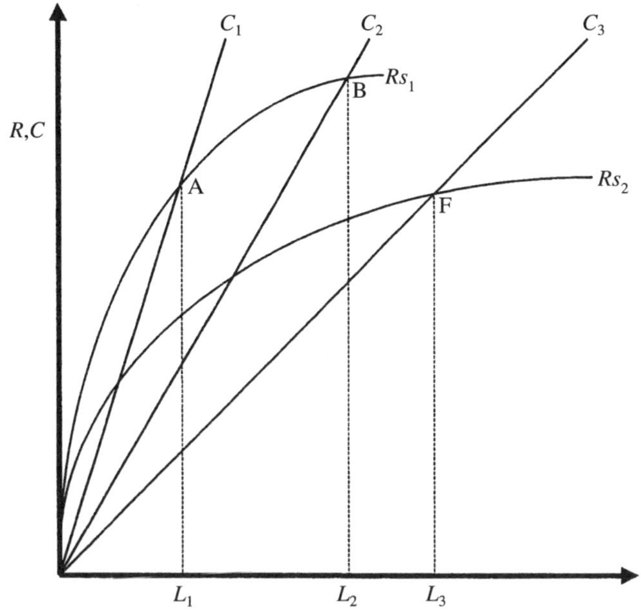

Figure 7.3 Revenue maximisation with given *L*.

Toposequences of land use and land degradation

A specific dimension of the population – land degradation nexus, revealed by many case studies, is the existence of a toposequential pattern of land use and the concentration of degradation processes on the uplands. Pingali and Binswanger (1987) find that population pressure leads to a sharp change in preference and price of different types of land (except in arid zones), with a movement to heavier bottom-land soils that are more responsive to intensification inputs but require high labour inputs for drainage and water control.

If, however, low-lying land becomes scarce, and cultivation is resumed in sloping lands, sustainable use requires investments in erosion control. Pingali (1989) observes that in the mid-slopes the return to such investments may be sufficient to induce private initiative; however he quotes case-study evidence showing that 'the probability that erosion control investments will not be made is higher in areas without well defined property rights and where group action is required for . . . watershed level investments'. On the upper slopes of the toposequence low returns to erosion control investments and the difficulty of defining rights over forest resources are more likely to result in degradation, and ultimately in low-yield equilibria with migration.

Some studies on Burkina Faso (Stoop 1987; Vierich and Stoop 1990) describe a similar toposequential pattern of land use: a first stage of long-fallow land use, mainly in upper slopes and plateaux; a second stage where declining fertility of the uplands induces the cultivation of river valleys; a third stage in which increasing population pressure forces farmers to expand cultivation on previously abandoned uplands. Degradation processes have been concentrated on these uplands, owing to their low initial fertility.

Similar conclusions emerge from a study on Rwanda. Kangasniemi and Reardon (1997) investigate whether increasing land scarcity, reflected in miniaturisation of farms, is associated with unsustainable land uses.[66] They find that the degree of protective crop cover was almost the same on small and on large farms, in three out of five agroclimatic zones. However, agricultural land uses appeared to be more erosive than elsewhere (particularly in small farms) at high altitudes. At such altitudes, which cover roughly one quarter of Rwanda's cultivable land, land scarcity was strongly associated with erosive land use practices.

World-wide evidence from the Global Assessment of Soil Degradation (GLASOD) project[67]

World-wide evidence on land degradation is available from the Global Assessment of Soil Degradation (GLASOD) project, on a cross-section basis (see appendix to this chapter). These data are used here in combination with other data bases to investigate whether they show any relationship between population pressure on land – as measured by population density – and land degradation, and to control the relationship for agroclimatic zones and other circumstances.

A traditional chi-square analysis of the data indicates that, as expected, severity of land degradation and population density are not independent. The data show that the relation is positive at least until certain threshold population densities. The coefficient of contingency – a standardised chi-square statistic – is highest between severity of land degradation and population density (0.497), followed by climate (0.400) and income group (0.365). Repeating this analysis by income group, in the case of the lower-middle-income economies the coefficient reaches a value of 0.640 for the association between land degradation and population density.

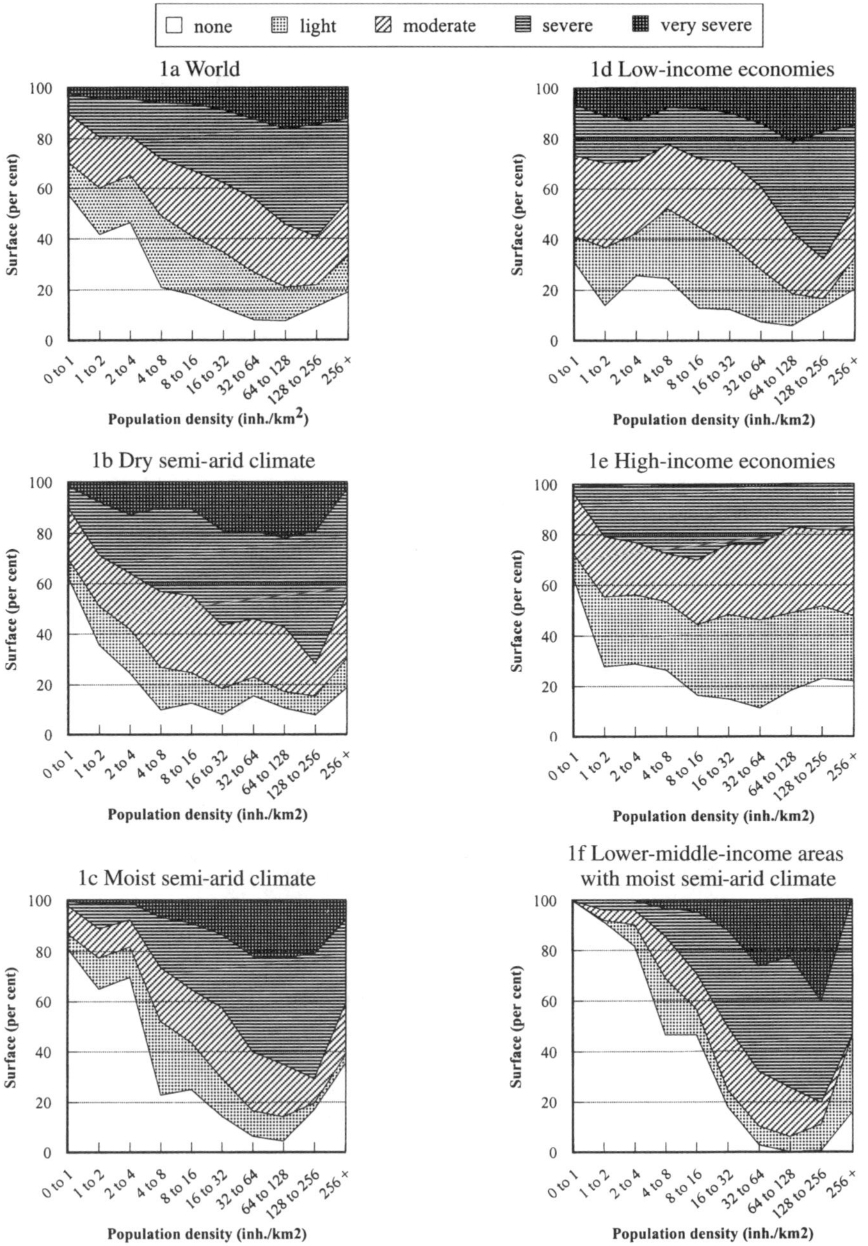

Figure 7.4 Population density and severity of soil degradation.

Figure 7.4 reports a selection of results. Figure 7.4a shows the relationship between population density and severity of land degradation at world level. For each class of population density (*x*-axis) the *y*-axis indicates the incidence of the various classes of severity. Thus, in areas with average population density in the range 0–1 inhabitants per square kilometre, one can see that 57.6 per cent of land shows no degradation, 12.9 per cent shows light degradation, 19.4 per cent moderate and 10.2 per cent severe and very severe. It appears that the incidence of all classes of degradation tends to increase with density, whilst a reversal is observed for the highest densities.

When units are disaggregated on the basis of agricultural potential, using the length of growing period as an indicator, a pattern of relationship with severity first increasing and then decreasing with population density appears to be generally confirmed. And, as may be expected, arid areas show a higher incidence of severity also at low densities.

The analysis by income group (low income, lower middle, upper middle, high) indicates that in high-income economies there are hardly any areas with very severe degradation. In general, when severe and very severe degradation are considered, there appears to be a very substantial decrease of the incidence of degradation (for all densities) between developing and developed countries.

This result confirms the observations on the relationship between poverty and failures of adjustment, although it must interpreted with caution, since income influences the relationship between land degradation and population density in two additional ways. First, it is negatively associated with the degree of dependence of people on agriculture and land products in general. Second, at the level of aggregation used here, there is a strong negative association between income levels and population growth rates in recent decades: population, and therefore density, has grown much faster in low-income countries.

Finally, a further disaggregation of units into areas with the same length of growing period and income shows that in low-income countries the incidence of degradation tends to be high and uniform for all classes of density in areas with unfavourable climate, while an increasingly clear positive association between density and severity of land degradation emerges as units with more favourable climate are considered. Most units present also a reversal of this association at the highest densities. In lower-middle-income economies, more than forty per cent of the observations are in the moist-semi-arid category, which shows clearly (Figure 7.4f) the U-shaped pattern of association between population density and severity of land degradation suggested by some of the case-study evidence.

In summary, these data appear to confirm some of the findings emerging from the case-study literature, namely that the severity of land degradation increases with density up to a threshold level, and that poverty (low income) and unfavourable climatic conditions are associated with more degradation at all density levels.

Appendix

The analysis is based on four sets of data: land degradation data from the GLASOD project (Oldeman *et al.* 1991); population estimates from the Global Demography Project (Tobler *et al.* 1995); FAO data on length of growing period (i.e. the number of days in a year when both moisture and temperature conditions are favourable for crop growth).

GLASOD was implemented by the International Soil Reference and Information Centre – ISRIC – (in cooperation with the Winand Staring Centre for Integrated Land, Soil and Water Research, the International Society of Soil Science, FAO and the International Institute for Aerospace Survey and Earth Sciences) for the United Nations Environment Programme. The project defines land degradation as the lowering of the current and/or future capacity of the soil to support human life – through water and wind erosion, chemical deterioration, physical deterioration – linked to a number of human activities, namely deforestation and removal of natural vegetation, overgrazing, agricultural activities, overexploitation of vegetation for domestic use and (bio)industrial activities (degradation due to this last type of activity has been excluded from the analysis, since it is not linked to agriculture. Its incidence on the results was however negligible). The severity of degradation is a combination of the degree and relative extent of degradation, which results in four severity classes (light, moderate, severe and very severe).

The data on land degradation, population and length of growing period are organised on the basis of a common geographic grid, providing an information set with approximately two million observation units for the entire world. The grid to which these data refer are 5-minute quadrilaterals. Due to the shape of the earth, the surface of the observation units varies considerably (between approximately 27 and 85 km^2), with its maximum at the equator and diminishing toward the poles – the highest latitudes included are 70°. In order to refer to land surface each observation unit in the database has been weighted with its respective estimated surface.

The fourth set of data is per capita GNP at country level, as estimated by the World Bank. These data have been used to classify each observation unit according to the income group of the country to which it belongs, since income data are not available for these geographic units. This is of course a gross simplification, but still it was considered preferable to ignoring the income variable.

All units are grouped according to ten classes of population density, which are formed according to a geometric scale. Density here refers to total population as data on population in agriculture are not available at this level of geographical disaggregation. Furthermore, the use of overall density is not inappropriate, since the causes of degradation considered include, besides agricultural activity, also deforestation and removal of natural vegetation, overgrazing, and over-exploitation of vegetation for domestic use; and since, for a range of densities, especially in poor countries, the dependence of population on agriculture, and generally on land products, is very high. One must remark, however, that such dependence varies with income: at high-income levels the dependence of population on agriculture and land products becomes very limited. This must be taken into account when interpreting the impact of the income variable on the relationship between population and land degradation.

Units were also divided into six classes of climate or 'length of growing period' (arid, dry semi-arid, moist semi-arid, subhumid, humid, perihumid), providing an indication of the extent to which the agricultural potential of each area is constrained by climatic conditions.

8 Future challenges for world agriculture

The new research challenges

The discussion in this book has concentrated on agricultural intensification in response to population growth, and the green revolution has been described as the most important recent and successful response. In Asia food production has more than kept pace with demand over the past thirty years and in many areas grain yields are among the highest ever achieved. Without the green revolution such supply increase would not have been possible and supply expansion would have occurred at the cost of opening new lands and depleting the soils.

The types of failures discussed so far refer essentially to the inability to intensify agricultural production while maintaining long-term soil productivity through the use of modern inputs.

However, there are also difficulties and environmental problems associated with the green revolution type of intensification. Research on improved plant varieties can be classified in three groups: research focused on raising yields at experiment stations and affecting the technological yield frontier; research on closing the gap between average farm yields and experiment stations' yields; productivity maintenance research – such as the research that is necessary to maintain pest resistance in the face of evolving pest biotypes. In the case of rice, for instance, after the realisation in 1966 of IR-8 – the first widely adopted modern rice variety – many new varieties had to be successively developed to replace those that began to give lower yields because of vulnerability to evolving pests and diseases.

Recent reflections on the green revolution seem to indicate that the yield gap between experimental stations and farms is closing, as more and more farms tend to operate at a point close to B in Figure 5.1. Furthermore, there is evidence that there may be a stagnant or even declining technological yield frontier, as there have been no breaks through the rice yield ceiling set by IR-8 in fertiliser response experiments. In terms of Figure 5.1 this means that no major shift seems to have occurred in the response function from u_1 throughout the era of the green revolution and that, unless new technologies are created, yields will eventually stagnate. The increase in average rice growth for South and Southeast Asia from 1.8 per cent per year during the 1970s to 2.8 per cent during the 1980s was due to the expansion of modern varieties to less-favourable production environments and to the development and diffusion of varieties more resistant to pests and diseases but not necessarily higher yielding than earlier MVs under ideal rice growing conditions (Hayami and Otsuka 1994).

More disturbingly, Pingali (1997) suggests that the technological yield frontier may be declining because of a degrading paddy microenvironment resulting from intensive rice monoculture. The author argues that such degradation may be due to a number of causes. Rice production has been subject to increased pest pressure, attributed to the uniformity of varieties grown, indiscriminate pesticide use, and increasing susceptibility of varieties to resistance breakdown. Relatively minor pests have caused noticeable losses as the area planted to modern varieties has increased. Furthermore, there is evidence of rapid depletion of soil micronutrients and changes in soil chemistry brought about by intensive cropping or increased reliance on low-quality irrigation water. In summary Pingali holds that: 'There is a growing concern about the resilience of the humid tropical lowlands and an increasing understanding that they do not have an unlimited absorptive capacity' (Pingali 1997: 225–226).

Table 8.1 shows the projected growth rates of population (UN medium variant), aggregate demand and production as estimated by FAO (2000).

The difference between the past growth rates of demand – shown in Table 4.1 – and these projections is nearly equal to that of the population growth rates. The deceleration is essentially explained by trends in a group of developing countries – including many of the largest in terms of population (China, Indonesia, Brazil, Mexico, Nigeria, Egypt, Iran, Turkey) – that started in 1995/1997 with fairly high per capita food consumption (over 2700 kcal/person/day) and are experiencing a significant slowdown in their population growth. In contrast, according to the FAO, in the other developing countries, including India, demand growth will decelerate less than population.

Hence, a global slowdown of demand for agricultural products will coexist with historically high growth rates in many areas of the world,

Table 8.1 Projected growth rates of population, aggregate demand and production (per cent p.a.)

	1995/7–2015			*2015–2030*		
	Pop.	*Dem.*	*Prod.*	*Pop.*	*Dem.*	*Prod.*
World	1.2	1.6	1.6	0.8	1.3	1.3
Developing countries	1.4	2.2	2.1	1.0	1.7	1.6
excluding China	1.6	2.3	2.2	1.2	1.9	1.8
sub-Saharan Africa	2.4	2.8	2.6	2.0	2.5	2.4
Latin America and Caribbean	1.4	2.0	2.0	0.9	1.5	1.6
South Asia	1.5	2.6	2.5	1.0	2.1	2.1
East Asia	0.9	1.9	1.8	0.5	1.3	1.2
Industrial countries	0.3	0.6	0.8	0.1	0.4	0.6
Transition economies	0.0	0.9	1.1	–0.1	0.6	0.6

Source: FAO (2000)

especially sub-Saharan Africa. It must also be stressed that these projected growth rates of effective demand and of production will reduce but not eliminate the problem of inadequate food intake: FAO (2000) estimates that the number of undernourished in developing countries will still be 400 million in 2030. At the same time, the possibility of expanding cultivation to new lands no longer exists in many areas of the developing world, while in others such expansion could only occur at the expense of forests.

There is consensus about the fact that future demand growth will have to be met essentially by increasing the productivity of land already in cultivation, i.e. through yield increases (Pinstrup-Andersen *et al.* 1999).

However, as Ruttan (1997) remarks, in general the sources of future productivity growth in agriculture are not as apparent or as attainable now as they were a quarter of a century ago.

> Incremental responses to increases in fertiliser use have declined. Expansion of irrigated area has become more costly. Maintenance research . . . is rising as a share of total research. The institutional capacity to respond to these concerns is limited, even in the countries with the most effective national agricultural research and extension systems. Indeed, during the 1980s, there was

> considerable difficulty in many developing countries in maintaining the agricultural research capacity that had been established in the 1960s and 1970s.
>
> (Ruttan 1997: 24)

Furthermore, it is not enough to know that food production could be increased globally, to match demand growth. Because the income of millions of poor depend on agriculture, there is a direct relationship between their ability to obtain food and agricultural productivity. In other words, the challenge is to increase productivity in the backward countries or areas, i.e. in those areas that were generally bypassed by the green revolution.

A further factor to be considered is the increased scarcity value of water that can be expected in the future. Agriculture is the dominant water user on a global scale; in the case of most of Asia rice is the largest user of water. At present, over-extraction of groundwater causes a range of environmental problems in some countries. In the future, increasing competition from domestic and industrial uses of water may result in increasing prices, with consequences difficult to predict (Alexandratos 1995a, b; Dawe 1998).

Finally, intensive agricultural systems, including the green revolution type of systems, have well-known and significant environmental spillovers. Chemical pesticides for instance have negative externalities that include air, soil and water pollution, elimination of beneficial organisms and the development of species resistant to the specific pesticides used and such external diseconomies are increasingly recognised.

On the whole, increasing the food supply for a growing population at an acceptable environmental cost will continue to pose a formidable challenge. Advances in basic science, especially molecular biology and biochemistry, open a host of new possibilities. However, the exploitation of such advances creates a new set of institutional issues, for the developed and especially for the less developed countries.

Biotechnology

In principle biotechnology[68] could provide answers to many of the problems discussed in relation to agricultural intensification. Innovations such as the development of an insect- or herbicide-resistant plant variety[69] can be represented as a non-neutral shift of the technological frontier. In the case of Bt varieties (Figure 8.1) the

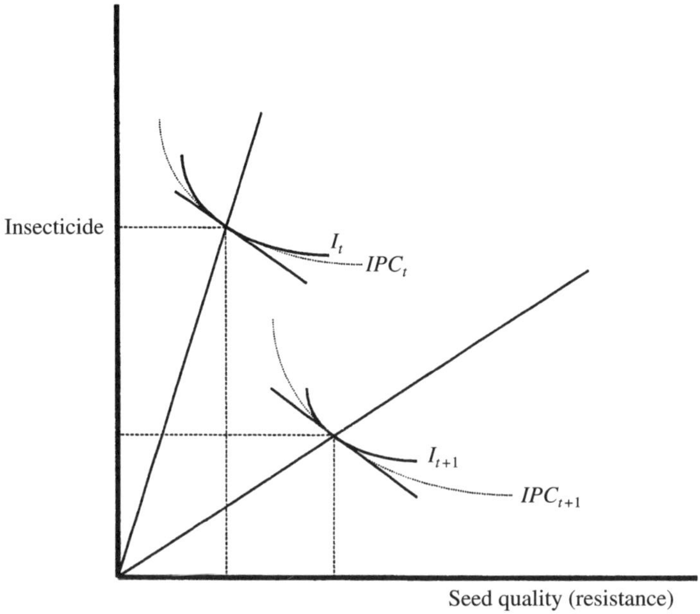

Figure 8.1 Bt varieties.

technology substitutes a conventional chemical input (insecticide) with seeds incorporating insect resistance.

In the case of RR varieties, seeds incorporating resistance to Roundup herbicide substitute for labour and other herbicide management inputs when used in combination with Roundup.

These shifts are produced by the combination of two developments: rapid advances in molecular biology and genetic engineering and the establishment of intellectual property rights (IPR) on innovations incorporated into seeds.

Historically, with the significant exception of hybrid corn in the USA, farmers' free access to plant varieties has been the rule in agriculture. In the USA, although legislation protecting plant breeders' IPR had been introduced since the 1930s, decisive steps in this direction have been linked to modern biotechnology, with the granting of utility patents encompassing living organisms. Specifically, a turning point was represented by the 1980 ruling of the US Supreme Court, in the much-quoted *Diamond v. Chakrabaty* case.

In the context of limited experience with patent issues in a field

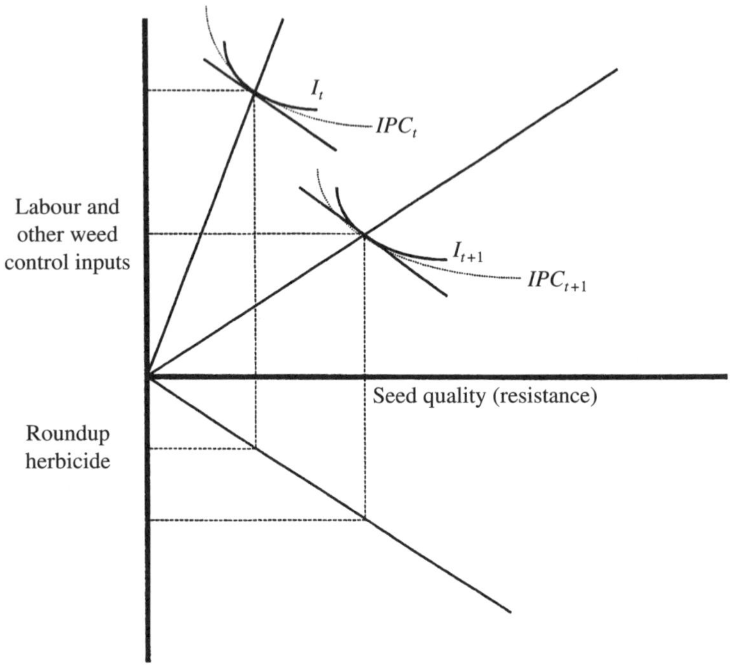

Figure 8.2 RR varieties.

entirely new such as biotechnology, the US Court ruled in a five to four vote that a live, human-made microorganism was patentable. What followed later – a pressure to maximise the number of patents – has been described as the closing of a global commons – represented by the plant genetic materials (Herdt 1999) – and as very much akin to speculative land grabs (Falcon 2000).

The complexity that this process has generated in terms of intellectual property rights is illustrated by the case of vitamin A rice, which in 2000 was reportedly based on seventy patents, held by thirty-two different firms (Falcon 2000).

Furthermore, biotechnology has allowed the development of many techniques designed to protect property rights. Important examples are RR varieties – where trait expression is linked to the use of a specific herbicide provided by the owner of the trait – and genetic use restriction techniques, known as 'terminator genes', that allow perfect property rights enforcement on seeds, as plants produce seeds that are sterile and must therefore be purchased each season.

Firms carry R&D and offer the new products because with intellectual property rights protection they can appropriate the benefits. The new varieties change the structure of costs for farmers – they reduce some variable costs (e.g. insecticide costs) and increase seed costs; farmers in the developed and developing countries adopt the new technology if it is, on the whole, cost reducing and pay a price premium for the modified seeds.

Biotechnology does substantially reopen the technology gap in agricultural innovations between the developed and the developing countries. As compared to traditional plant breeding it allows responses that are faster (the importance of 'time' has been extensively discussed in this book), more flexible (i.e. the innovations can fit the needs of many different environments, therefore do not require the set of conditions that limited the success of green revolution varieties to certain regions of the world), more precise (i.e. they produce varieties whose traits correspond more precisely to the research objectives than those obtained through conventional plant breeding)[70] and have the potential to address some of the externality problems of agricultural intensification.

Plant breeding has evolved over time exhibiting a dramatic increase in the speed with which new varieties were produced, from the several generations required for developing a new variety in the context of informal innovation by farmers, to the great acceleration impressed by scientific plant breeding during the last century. Modern biotechnology represents a quantum leap in this process. For example, molecular marker technology can halve the time needed to produce new varieties with resistance to important crop diseases. The same technique accelerates developments as varied as pest control traits (like genetic resistance to insects); agronomic traits such as tolerance to drought, salinity and heat; post-harvest traits such as delayed ripening of fruits; output traits, such as specific food qualities. Flexibility is increased because transgenic approaches allow the introduction into crop plants of genes from any class of organisms and therefore considerably broaden the range of gene pools that can be used for crop improvement purposes (Lindner 1999; Maredia *et al.* 1999).

As for the issues that have dominated the public debate, it should be recalled that the possible negative side effects of biotechnology[71] are in many ways analogous to the type of externalities associated with intensive agriculture and it is not clear whether, on balance, problems such as pest resistance and loss of biodiversity may be reduced or amplified with the new technologies. However, while the negative externalities of agricultural practices, such as chemical inputs use, are known, the side effects of the genetically modified (GM) crops are surrounded by a higher degree of uncertainty, because biotechnology allows genetic crossing among very distant classes of organisms. Uncertainty and the related consumers' diffidence raise the institutional costs of innovations both for the private and for the public sector, since for instance, the safety assessment on GM crops must be more careful than with the outputs of conventional plant breeding (Dale 2000).

Given the great potential of biotechnology, one could imagine seeds that eventually provide answers to most of the problems of poor farmers (low-input agriculture) in unfavourable environments and therefore to most of the problems of intensification discussed in this book. These 'miracle' seeds could combine increased resistance to pests and to environmental stresses such as drought, and improved nutritional quality. Furthermore, the embodiment of desirable qualities into seeds would reduce both the human capital requirements of farming at the farm level and farmers' dependence on input markets (other than the seed market), both desirable qualities.

In practice however, the impact of biotechnology on the problems of agricultural intensification discussed so far are largely dependent on the evolution of a number of market structure and institutional issues.

Biotechnology is closely linked to the establishment of intellectual property rights in agricultural innovations – in contrast with the tradition of sharing of genetic materials between public research centres engaged in plant breeding. The establishment of IPR should foster innovation to the extent that it provides private incentives for R&D. The historical experience of hybrid maize[72] in the USA, and other more recent examples, actually show that IPR definition is associated with very high rates of private R&D investments and of varietal development.

If biotechnology translated solely into greater private investments it would unambiguously contribute to growth. However, past experience shows that IPR definition resulted in a 'crowding out' of public spending in plant breeding; furthermore biotechnology has been associated with industry concentration.

The sector has rapidly moved from a competitive structure with a large number of small innovative firms, to an industry dominated by six large multinational chemical and pharmaceutical corporations (Monsanto for instance was a traditional chemical company) through a process of mergers, joint ventures and acquisitions directed first at the smaller biotech research firms, later at the seed companies.

One possible set of explanations for this trend is related to the very process of IPR definition: both to its nature and to its still unsettled status.

The definition of intellectual property rights on innovations typically tries to achieve the right balance between providing incentive to discoveries (granting the innovator the right to prevent their use by others without permission) and insuring the technology transfer (e.g. by limiting in various ways the breadth of patents and imposing that the invention is described in a way that permits duplication on patent expiration).

However, the variety and reach of IPR claims related to biotechnology processes and inventions, including those that represent enabling technologies for research, has a number of consequences: (i) It is a direct source of monopoly power. (ii) It entails a substantial amount of litigation, i.e. substantial transaction costs. This in turn is an incentive to concentration. Firms would acquire related companies in an effort to reduce contractual hazards, by internalising IPR disputes. (iii) The 'gridlock' of patent rights could slow down the pace of innovations, although the biotechnology companies try to ensure their 'freedom to operate' (even while engaging in litigation in the courts) by reaching explicit or tacit agreements on cross-licensing (Lindner 1999). Finally, the necessity to seek and defend intellectual property rights over key technologies and a complex system of regulations impose high fixed institutional costs, which represent a further incentive to concentration.

There are fears that lack of competitive pressure may eventually result in a reduction of the pace at which innovation proceeds and of the choices available to farmers and consumers.

As for the role of the public sector, the experience with hybrid maize in the USA shows that the protection of property rights results in very high rates of investment and variety development and in the dominance of private seed companies.

The substitution of public with private plant breeders is an indication of the latter having a competitive advantage over the former. However, in the case of biotechnology, there are also indications that the public sector may have some disadvantages *vis-à-vis* large biotechnology companies in defending its 'freedom to operate'. Tentative explanations include the fact that its internal decision mechanisms make it costly to negotiate cross-licensing and other agreements, and the fact that public institution would not sue for anti-competitive behaviour (Lindner 1999).

Furthermore, the public research centres have experienced for decades the consequences of financial restraint and have been under pressure to become partially self-financed and to substitute public with industry-provided funds. Biotechnology has increased these pressures.[73]

On the whole, it seems clear that if the public sector has to retain a distinct role in biotechnology a substantial institutional response is required at the national level and in the international institutions.

From the developing countries' point of view there are a number of issues. The private life-science companies in the developed countries increasingly define the research priorities in agriculture. The private sector is guided by the profit motive and therefore by the extent of the effective demand for innovations, i.e. by the extent of the actual or potential value of the market for seeds. Therefore research includes all the major food crops;[74] the 'life-science' companies should be expected to try to introduce available products (innovations) in many developing countries and to address directly specific developing countries' problems that involve large numbers of non-poor farmers.[75]

However, the discussion of agricultural intensification in response to population growth has indicated that most of the historical failures have been related to poverty and/or very unfavourable environments.

The low input–low yield equilibrium of smallholder/peasant farming systems in many parts of the developing world is the result of a number of constraints: cash and credit constraints limit the possibility to acquire inputs; poor infrastructure in terms of input supplier networks means that input deliveries may be late and inputs may be of poor quality;[76] poor infrastructure in terms of storage and roads means that output price fluctuation tends to be locally wide.

Farmers in these systems use carry-over-planting material. With biotechnology they should eventually buy seeds, but this may happen only if those seeds allow them to overcome some of their economic and natural constraints. The experience with hybrid maize – which requires all the inputs (purchased seeds, fertiliser, pesticides and good rain) at the appropriate time – shows that an innovation which does not address those constraints has a low probability of being adopted.

The public research systems in the developing countries and the international research system have a role in addressing the demand for innovations of poor farmers and the safety concerns of consumers in the 'backward' countries.

With biotechnology the increasingly private-good nature of innovations will not allow the type of transfer that has taken place with the green revolution, i.e. it cannot be assumed, like in the induced innovation model, that the IPC be the same world-wide, or it should be expected that the time lag with which the 'frontier' becomes available at no cost to the national research systems in the 'backward' countries becomes larger. In the meantime adopting farmers in the 'backward' country pay rents to IPR that belong to foreign seed companies.

Falcon (2000) recalls the story of the wheat dwarf varieties – that was brought to the USA in the 1930s, reached CIMMYT in the 1950s and forty years later represented more than 80 per cent of the wheat grown in the developing countries – and suggests that this story most likely could not be replicated with biotechnology in the current IPR environment. Private firms would have little incentive to invest in research for a self-pollinated crop, generating little seed sales; they could have developed some hybrid lines, but the extent and speed of dissemination would most likely have been lower (Goeschl and Swanson 2001).

With the exception of a few large countries (including India, China, Mexico and Brazil) developing countries have little research capacity and no regulatory framework to legally acquire and safely release biotechnology products (Maredia *et al.* 1999). For example, a recent survey on the management of proprietary technologies within the CGIAR and a group of Latin American national research organisations (Komen *et al.* 2000) found that the ability to manage IPR (ability to obtain clear knowledge or information regarding the type of IPR established for a particular innovation and the associated restrictions in terms of research and post-research use, ability to seek intellectual property protection for own inventions) was limited both in the CGIAR and in the national research organisation studied.

Furthermore, it appears that the role of the private sector in terms of biotechnology R&D is still very limited.

Maredia *et al.* (1999) list two reasons why biotechnology actually poses problems of public investment decisions in the developing countries. One is the opportunity to exploit the technology spillovers – from the developed countries R&D sector – that may easily occur at the process level. The other is the difficulty to develop a substantial private involvement in biotechnology R&D in countries where the size of the market is small (because farmers are poor) and/or IPR enforcement is too expensive (because small farms are prevalent). They hold that 'For most self-pollinated crops in small-farm agriculture, the public sector, both international or national, will continue to play the leading role' (Maredia *et al.* 1999: 7).

Biotechnology requires at least some type of investment in high-skills human capital (in the biological and legal fields) and regulatory frameworks. One may argue that some of these adjustments are no more challenging than those required by the green revolution. For example, it is no less difficult to develop water control in a vast countryside, involving large numbers of farmers and local institutions than to 'produce' PhD students and regulatory frameworks. Both can be obtained on the basis of the institutional build-up in the developed countries, presumably at very low cost. Also, the public research institutions in the developing countries could improve their capability to exploit complementarities and economies of size at the national level and through regional cooperation. On the other hand there may be costs associated with the fact of being 'policy takers' in terms of regulatory frameworks for IPR and safety and the implementation of such frameworks may be difficult.

In conclusion, biotechnology has an enormous potential but poses a set of issues of regulation and institution building, both in the developed and in the developing countries. Although the challenge is formidable, one must recall that there may be little technological alternative and that also the success of the green revolution was linked to a set of institutional preconditions.

Conclusions

The assessment of the links between population and economic development requires the consideration of the direct and indirect impact of population on the economy and on institutions and of economic growth on population. While analytical models have gone a long way in explaining the impact of economic variables on demographic change, economic models tend to treat demographic variables as given and a consensus model of the complexity of the interaction has certainly not emerged.

Both theory and empirical evidence suggest that when population grows neither the Malthusian outcomes, nor the 'capital shallowing' effects of the standard neoclassical model and the 'tragedy of the commons' are inescapable. At the same time simple functionalism and the invisible hand metaphor are inadequate instruments to deal with the complexity of the actual population problem. Even if societies tend to produce the right responses when the underlying economic conditions change, the starting conditions – especially the initial distribution of assets – and the pace at which such changes occur are critical. In the case of the present developing countries the pace imposed by the rates of population growth experienced since the 1950s is historically unprecedented.

Within the general debate on the linkages between population growth and economic development, references to agriculture are often used as examples of the ability of societies to produce the right economic and institutional responses to demographic shifts. The fact that the prices of major staples have declined in the context of population growth can be taken as prima facie evidence of such ability. Such a view is also supported by the models of endogenous, population-driven, technical progress and of induced, factor-saving innovations, as well as by the theory of the evolution of property rights on land and by a related body of empirical evidence on contemporary less developed countries.

The discussion in this book suggests a less optimistic picture. Population growth actually induces adjustments that on the whole make it possible to meet growing food needs, and to maintain agricultural incomes (although much less successfully) and land quality. However, optimism about such adjustment mechanisms must be tempered when the role of inequality in the distribution of assets is considered, the environmental resource base is taken into account and the possibility of land degradation is introduced.

The unequal distribution of assets, most importantly land, may interfere with the success of collective action, reduce yield growth and expose some groups – the landless and near landless – to a condition of inadequate food intake. With rapid population growth, in the context of poverty, inequality and unfavourable climate, communities may proceed too far along a given production function causing a reduction of long-term land productivity, before a new technology is introduced. Furthermore, intensive agricultural systems have well-known and significant environmental spillovers, including surface and groundwater contamination from plant nutrients and pesticides.

In spite of the perspective of a slowdown in the pace of growth of population and demand in many developing regions, the demand for agricultural products is likely to continue to grow at rates that are historically high. At the same time the possibility of expanding cultivation to new lands no longer exists in many areas of the developing world, while in others such expansion could only occur at the expense of forests. Therefore future demand growth will have to be met essentially by increasing the productivity of land already in cultivation.

In many parts of the developing world the demand of a rapidly growing population has been met by introducing high-yielding varieties of cereals. This has been accomplished through a massive public investment in the transfer of technology from developed to developing countries – the green revolution.

The 'miracle seeds' of the green revolution however cannot provide all the answers to rapid population growth in less favoured areas and in the already intensive systems. The successful introduction of high-yielding varieties has a number of preconditions, chiefly an adequate irrigation infrastructure, and this in turn entails a problem of institution building. This tends to be a slow process, especially where the distribution of land assets is very unequal. As for the already intensive systems, the possibility of further yield increases based on the heavy use of chemical inputs and of water has negative environmental effects that cause increasing concern.

As for the promises of biotechnology, they are indeed enormous, but the time lag that separates research from its practical impact on yields and the environment is usually large and so are the requirements in terms of institutional adaptation, especially for the less developed countries. In the latter, investments needed to build even a minimum level of biotechnology research capacity include investments in research capacity, in the capacity to review and manage the environmental and human safety implications of GM crops, and in a regulatory framework for IPR protection.

In summary, agriculture in contemporary less developed countries illustrates well both the merits and the limitations of a 'revisionist' approach to the population problem and of a functional view of institutional adjustments.

Although the Malthusian outcomes have proved avoidable, many local failures occur, and the environmental resource base is often affected. As for the future, the possibility of increasing the food supply for a growing population at an acceptable environmental cost will continue to pose a formidable challenge.

Notes

1 For a discussion see Livi Bacci (1997).
2 'L'histoire immobile', *Annales E.S.C., XXIX* (1974: 689), as quoted in Aston and Philpin (1988).
3 The relation between the epidemics and sociodemographic conditions is a controversial topic. Some authors (Boserup 1981; Livi Bacci 1997) hold that the plague acted as a largely exogenous population check.
4 Interestingly, the observations of historians suggest that a famine could hit the poor even without a serious shortage (Genicot 1971: 810), a fact that has been analysed by Amartya Sen (1981) for contemporary developing countries.
5 Postan contends that – in spite of the legal codification of villeins' submission in the twelfth and thirteenth centuries – the main weight of their disabilities was economic. For the villeins, customary rent was considerably higher than that of freeholders and a number of freeman rights had to be eventually bought by the villein.
6 Kelley (1988) reports a Solow commentary to the Forrester–Meadows model in a contribution entitled 'Is the End of the World at Hand?' (*Challenge*, March–April 1973: 43). 'the characteristic conclusion is very near the surface. The basic assumptions are that stocks . . . are finite, that the world economy tends to consume that stock at an increasing rate . . . and that there are no built-in mechanisms by which approaching exhaustion tends to turn off consumption gradually and in advance. You hardly need a giant computer to tell you that a system with those behaviour rules is going to bounce off its ceiling and collapse.'
7 The authors also observe that, even if one accepted the alternative hypothesis that the obstacle is a deficiency of effective demand, population growth would not necessarily be positive, as it may not translate into growing effective demand.
8 This discussion of the model is based on Solow (1956, 1987), Barro and Sala-i-Martin (1995) and Valdès (1999). The functional dependence of variables on time in growth models is henceforth left implicit.
9 Constant returns to scale imply that if K and L are multiplied by a scalar λ the level of output will also multiply by the same scalar $F(\lambda K, \lambda L) = \lambda F(K, L)$.
10 $$\frac{\mathrm{d}k}{\mathrm{d}t} = \frac{[L(\mathrm{d}K/\mathrm{d}t) - K(L_0 n\mathrm{e}^{nt})]}{L^2} = \frac{\dot{K}}{L} - nk$$

11 For a formal demonstration of this result see Barro and Sala-i-Martin (1995).

12 The constancy of the capital–output ratio and of the rate of profit in the long run (or, equivalently, a constant functional distribution of income) is one of the 'stylised facts of growth'. There is a mathematical proof (although not a clear economic intuition for it) that such constancy is ensured – as technical progress takes place – only if technical progress is labour augmenting. This type of technical progress is also called 'Harrod neutral' (Harrod defined technical progress as neutral when the capital–output ratio and the marginal productivity of capital remain constant, which, in a competitive economy, implies that the relative income shares remain constant) (Valdés 1999).

13 Kelly and Schmidt (1994) point out, however, that more recent studies, covering the period up to 1985 show (weak) empirical evidence of a negative correlation between output and population growth. Their own estimates, extending to the late 1980s show a 'reasonably consistent' negative correlation, for the 1980s, while confirming a nil impact of population growth on per capita output growth in the 1960s and 1970s. Their explanation for this result is based on the decomposition of the net growth rate of population into its components: net mortality reductions and net births.

14 In Stiglitz's analysis sharecropping performs a role in sharing risks between landlords and tenants; because of high costs of monitoring labour inputs, it provides also incentives to work and, since it is a contract in which pay is based on performance, may also act as a self-selection device.

15 In the two fields system all arable land is ploughed but only half planted to crops, the other half being left fallow to recuperate its fertility.

16 In the archetype the system was based on the division of arable land into three parts. Each year there was one part planted with a winter crop (wheat); one planted with a spring crop such as oats, barley or legumes; one left fallow. The next year crops were rotated, the first field being left fallow, the second planted to winter crops, the third to spring crops, and so on.

17 A debate on the transition from feudalism which appeared in *Past and Present* between 1976 and 1982, two decades after the debate among Marxist historians on the transition from feudalism to capitalism which had started in 1950 in *Science and Society* (see Aston and Philpin 1988).

18 The terms common, communal and traditional are used indifferently to indicate tenure systems based on some form of group control on land, as distinct from purely private property and state property. This topic is discussed on pp. 93–109.

19 There are other possibilities: F may be an increasing function or, alternatively, returns on the CPR may diminish very sharply and $w > F(2)/2$. In this latter case there are two equilibria, each with one player investing in the CPR and the other outside the CPR.

20 A trigger strategy is a strategy under which a player can use two single-shot actions. The player will begin by choosing one of the two actions and will continue to choose it if all other players do the same but will switch immediately and permanently to the other action if she detects any other player doing so (Friedman 1990).

21 There are alternative solutions, including 'reputation': even a small probability that the player is of a type that prefers to cooperate acts as an

incentive to all other players to behave cooperatively. In a model (Seabright 1993) historical considerations – rather than expectations of future gains – explain successful collective action, as there is a tendency of cooperative behaviour to enhance the prospects for successful cooperation.

22 Mason and Fry, in a 1982 paper find that on balance, a rise in population growth above a certain level depresses the saving rate to extents that increase with the level of the growth rate of per capita income. One simulation of the saving ratio over the demographic transition from high to low fertility yields a pattern in which the saving ratio first increases and then returns to pre-transition levels (Mason 1988).

23 According the World Bank (1984) in the developing countries the poor, who are the majority, meet child rearing costs by reducing other members' consumption or by unsustainable use of resources such as land.

24 Countries with the same production function and the same values of the parameters s, n and d would have the same steady-state values k^* and y^*. Since the relation between the level of k and its growth rate is negative in the model, if the initial k values differed among countries – because of past specific events – poor economies should grow faster per capita than rich ones, converging to the common steady-state values. Although the debate on this issue is far from being conclusive, we may say that the empirical evidence accords better with another notion of convergence, the idea that an economy grows faster the further it is from its own steady state, as determined by differing values of the key parameters – i.e. 'conditional', rather than 'absolute' convergence (Barro and Sala-i-Martin 1995).

25 The presence of a positive externality in the production function has also the effect of generating a divergence between the private and social marginal productivity of capital and therefore generates a sub-optimal rate of growth of the decentralised economy.

26 Similarly the scale effect of L on growth can be eliminated in a model where learning and spillovers apply to human capital by assuming that each producer benefits from the average level of human capital in the economy, rather than from the aggregate of human capital (Barro and Sala-i-Martin 1995).

27 For β, γ, a_L and a_K equal to 0 and $\theta = 1$ the model simplifies to the Solow model with a constant, exogenous rate of growth of technology and Cobb–Douglas production function.

28 An increase in L in the model increases the demand faced by each monopolistic firm selling a durable good and raises the return to human capital employed in manufacturing at the same time that it raises the return to human capital in research. These two effects exactly cancel. Romer points out that this exact cancellation is not a robust feature of the model: if slightly different functional forms were used, the net effect of an increase in L could be either to increase or to reduce the amount of human capital used in the research sector.

29 In a less simplified version of the model, population – rather than adjusting instantaneously to its steady-state level – grows at finite speed when income is above its steady state, hence per capita income rises over time. The recent levelling off of population can then be obtained in the model by assuming that population increases in income at low levels of income and decreases in income at high levels of income.

30 A contrasting institutionalist interpretation of fertility decline is given by John Caldwell (1996).

31 An additional restriction on a is that $a < 1$, i.e. the weight attributed to each child's utility is never larger than the weight attributed to the marginal utility of own consumption.

32 The negative trends in the former centrally planned economies (transition economies) were linked to the exceptional political and economic circumstances that those countries experienced during the 1980s and 1990s.

33 According to FAO, the number of undernourished in the developing countries was 960 million in 1969/1971 and about 800 million in a 1999 assessment. Although this number implies a sharp decline in the relative incidence of undernourishment (from 37 to 18 per cent of the total population of the developing countries) it still represents a massive problem of inadequate food consumption.

34 In the context of a discussion of the appearance of urbanisation in the ancient world, Boserup (1981) suggests that urbanisation is linked to density in two ways: in the first place the survival of towns required the production of an adequate agricultural surplus; this in turn, given the state of the transportation, conservation and agricultural production technologies, required a minimum population density in the surrounding countryside. In the second place the construction of buildings and roads, given the state of technology, required a large workforce.

35 All developing countries of Africa, Asia and Latin America with a population of at least seven million inhabitants in 1993 are included. The data are 1961–1963 and 1991–1993 averages. The changes refer to the ratios between: active population in agriculture and agricultural land (x-axis in Figures 4.2 and 4.3); agricultural production at constant 1989–1991 international prices and agricultural land (y-axis in Figure 4.2); agricultural production at constant 1989–1991 prices and numbers of active in agriculture (y-axis in Figure 4.3).

36 In addition, Pingali and Binswanger (1987) find that improved access to markets through better roads and transport facilities also leads to more intensive systems of land use; and that population pressure leads to a sharp change in preference and price of different types of land (except in arid zones), with a movement to heavier bottom-land soils that are more responsive to intensification inputs but require high levels of labour for drainage and water control.

37 Boyce (1987) tried to separate cause from effect in the case of West Bengal and Bangladesh by testing whether rural population growth in 1931–1951 affected agricultural output in the 1949–1980 period – i.e. using population as the lagged independent variable. The results indicate that population growth had a positive effect upon agricultural growth. According to the author there is also evidence of a positive linkage running in the reverse direction, from agricultural growth to rural population growth, particularly in Bangladesh, but this should be mainly due to variations in net migration to rural areas in other districts and to urban areas (Boyce 1987).

38 Sen (1981) very effectively observes that the reason why there are no famines in the rich developed countries is not because people are richer on the average, but because of the social security system.

39 In a similar vein, Binswanger *et al.* (1989) describe the transition from low-

density, land-abundant communities, to high-density, land-scarce communities as a process which encompasses Boserupian induced innovation and investment, the transition from general to specific land rights, and a set of endogenous changes in production relationships.

40 Hayami and Ruttan explicitly acknowledge the influence of the work of Demsetz, Olson, Coase, Williamson, Cheung, and of the historical interpretation of North and Thomas.

41 Non-rivalry is a typical attribute of the research product; non-excludability often applies in the case of the product of biological technology, since the nature of the agricultural production process makes it difficult to restrict the use of new information.

42 Although the dynamics of adoption under a technological treadmill will eventually lead to the convergence of actual demand towards latent demand.

43 In addition the evidence shows unambiguously that the introduction of the high-yielding varieties widens the absolute income gap between the owners of land and capital and the landless labourers, as the latter receive a small share of the incremental value of production.

44 The International Rice Research Institute (IRRI) and the International Center for the Improvement of Maize and Wheat (CIMMYT) were established respectively in the Philippines and in Mexico around 1960 on the basis of cooperation between the Ford and Rockefeller foundations. During the 1960s other centres were created; several international organisations became involved in the funding and management of programmes and a coordinating mechanism was established with the creation of the Consultative Group on International Agricultural Research (CGIAR), whose initial membership included three foundations, the World Bank, the FAO, the UNDP and a number of national governments (Hayami and Ruttan 1985).

45 One reason is that, being shorter than traditional varieties, they can produce heavier yields without falling over.

46 In the case of China, village communities own the land, and farmers' households the right to use it.

47 In an area of northern Nigeria surveyed in C. Udry (1993) credit markets were almost completely segmented along geographic and kinship lines, in spite of the fact that the study area was an environment characterised by highly correlated risk and seasonal demand for finance. This suggests that the information costs of such transactions should be high.

48 Such policies were in fact among the main recommendations of a 1975 'Land Reform Policy Paper' establishing the guiding principles of the World Bank land policy.

49 This is a function of the probability of eviction after the investment and the likelihood of recovering all or part of the investment if evicted.

50 Demsetz (1967) concedes that these legal and moral experiments may be hit-and-miss procedures to some extent.

51 In turn, the system of property rights influences the demand and supply of children of the different social groups, as, for instance, privatisation of land weakens the insurance motivation for large family size for the landowners, while increasing landlessness may retard fertility decline, since agricultural workers have less motivation than landowners for fertility limitation (Boserup 1989; Stokes and Schutjer 1984).

52 Furthermore, the productivity differential in favour of small farms is larger where inequalities in the distribution of land holdings is greatest.

53 Although analyses based on total factor productivity generally confirm the inverse relationship.

54 A succinct, but fairly exhaustive discussion of this topic can be found in Ellis (1993).

55 At very low population densities large farms imported slaves as workers – as in the slave plantations of the USA south east; at medium population densities – as in medieval Europe, China and pre- and post-colonial America – farmers owned the lord labour and other tributes; at higher population densities large farms relied on indentured labour (Binswanger *et al.* 1995).

56 The term ‘rent seeking’ indicates the competition to capture the excess profits (rents) which can be artificially contrived typically through government regulation (e.g. a government granted monopoly right). This competition leads to expenditures (e.g. for lobbying activities) that create no value from a social point of view.

57 See for instance Agarwal (1994 and 1997).

58 An example of this type of problems, for Senegal, is given by Lele and Stone (1989): ‘The long standing antipathy between the Wolof, for example, who dominate the Groundnut Basin, and the Diola, a non-Muslim group inhabiting the lower Casamance, is likely to complicate migration in Senegal. If the relatively well watered Casamance is to become an agricultural growth area for Senegal, the Diola will have to be given a greater share of national resources and be represented in the elite’.

59 Seventeen of Africa’s forty-five colonies gained their independence in 1960 followed by the Portuguese colonies of Angola, Zimbabwe, Mozambique and Cape Verde in 1975.

60 Lele and Stone (1989) emphasise this latter point, by stating that one shortcoming of the Boserup hypothesis is that it is based on the assumption of a benign, or at least neutral, policy environment.

61 Alston *et al.* (1996) have shown how inconsistent policies led to widespread violence in the process of property rights definition on two Brazilian frontier areas.

62 The model is described in Sanderson (1994) who quotes the unpublished thesis of A. Picardi.

63 See for instance Myers’ (1991) analysis of the impact of the 1978 Land Use Act, in Nigeria.

64 A different process is described by Jodha (1987) in reference to a semi-arid area of India, where the 1952 agrarian reform and subsequent public policies, intended to provide land to the landless and smallholders, translated into a large-scale privatisation of marginal common-property land and the abolition of former regulatory measures. This resulted in a qualitative decline of residual common property and increased soil erosion and declining yields in the privatised, former common-property grazing land that was converted to crops.

65 A different view is expressed by Rao (1994), who argues that, since children are a drain on family incomes at least until age five or six, with plausible discount rates, they lower, not increase, expected incomes. Their employment as workers would therefore be a consequence, not a cause, of high dependency rates.

66 They calculate an index of protective vegetative cover provided by crops or other land uses, such as pasture, by farm size.
67 This section is based on Cuffaro and Heins (1998).
68 According to a 1989 definition by the Office of Technology Assessment of the US Congress, biotechnology is any technique that uses living organisms, or substances from those organisms, to make or modify a product, to improve plants or animals, or to develop micro-organisms for specific uses. Biotechnology includes a continuum of technologies, ranging from long established and widely used technologies, which are based on the commercial use of microbes and other living organisms, to the more strategic research on genetic engineering of plants and animals (Persley 1994: 374–375).
69 Herbicide-tolerant and insect-resistant plants are the most widely used agricultural biotechnology innovations. Bt varieties of maize and cotton incorporate a genetic sequence of a micro-organism (*Bacillus thuringiensis*) so that plants produce proteins toxic to certain species of insects. Monsanto's RR (Roundup-Ready) soybeans incorporate genes that permit the plant to tolerate a very effective herbicide (Roundup) previously developed by Monsanto.
70 In conventional plant breeding, improved varieties can be obtained either through hybridisation between plant species and genera or by causing genetic changes through radiation or chemical mutagens. In the first case undesirable genes are typically transferred together with the genes that confer the desired trait (e.g. pest resistance); in the second case the genetic mutation is random. In both cases it is typically necessary to carry out a lengthy selection process among the plant lines obtained (hundreds or thousands in the case of mutation breeding) to arrive at a variety with the desired characteristics (Dale 2000).
71 In the fields, Bt varieties could induce the evolution of Bt-resistant insects and also kill non-target insects, modifying the ecological equilibrium; insect and herbicide resistance could be transferred to weedy relatives. Globally it is feared that the role of multinationals in biotechnology innovations may determine a drift towards genetic uniformity and genetic diversity erosion. As for human health, the use of genes for antibiotic resistance in biotechnology raises worries about the possibility that such resistance may be transmitted to consumers of GMOs.
72 Property rights are enforced for hybrids because hybrid seeds, when replanted, lose productivity. Farmers must buy seeds each planting season.
73 In the USA the role of royalties, grants, contracts and donations from the private sector has substantially increased. Some universities have accepted equity stakes in private companies as payment for the transfer of technology (Zilberman *et al.* 1999).
74 Hence, crops that are specific to relatively small areas or poor agricultural environments (e.g. cassava and yams) have been termed 'orphan commodities'.
75 Traxler *et al.* (1999) quote as an example the development by Monsanto of a genetically modified variety of Bt soybean effective against an important pest in Brazil.
76 Private companies may sell pesticides without standard tests or controls, chemicals are often past their sell-by dates and in some instances bootleg

(false) compounds have been sold. Farmers have reasons not to trust the private (often newly privatised) companies (Robinson 2000). In discussing reasons for low adoption of maize hybrids in Ghana, Santaniello (2000) quotes the poor performance of the government-owned seed company in supplying farmers with quality seed in a timely fashion.

Bibliography

Agarwal, B. (1994) *A Field of One's Own: Gender and Land Rights in South Asia*, Cambridge: Cambridge University Press.

—— (1997) 'Gender, Environment and Poverty Interlinks: Regional Variations and Temporal Shifts in Rural India 1971–81', *World Development*, 25(1): 23–52.

Ahmad, S. (1966) 'On the Theory of Induced Innovation', *Economic Journal*, 76: 344–357.

Akerlof, G.A. (1984) *An Economic Theorist's Book of Tales*, Cambridge: Cambridge University Press.

Alchian, A. and Demsetz, H. (1973) 'The Property Rights Paradigm', *Journal of Economic History*, 33(1): 16–27.

Alexandratos, N. (ed.) (1995a) *World Agriculture: Towards 2010. An FAO study,* Chichester: FAO and J. Wiley.

—— (1995b) 'The Outlook for World Food and Agriculture to Year 2010', in N. Islam (ed.), *Population and Food in the Early Twenty First Century*, Washington DC: IFPRI.

Alston, L.J., Libecap, G.D. and Schneider, R. (1996) 'Violence and the assignment of property rights on two Brazilian frontiers', in M. Garfinkel and S. Skaperdas (eds), *The Political Economy of Conflict and Appropriation*, New York: Cambridge University Press.

Andrè, C. and Platteau, J.P. (1998) 'Land Relations under Unbearable Stress: Rwanda Caught in the Malthusian Trap', *Journal of Economic Behavior and Organization,* 34(1): 1–47.

Arrow, K.J. (1962) 'Economic Implications of Learning by Doing', *Review of Economic Studies*, 29(1): 155–173.

Arthur, B. (1988) 'Self-Reinforcing Mechanisms in Economics', in P. Anderson and K. Arrow (eds), *The Economy as an Evolving Complex System*, Reading MA: Addison-Wesley.

Aston, T.H. and Philpin, C.H.E. (1988) *The Brenner Debate. Agrarian Class Structure and Economic Development in Pre-Industrial Europe*, Cambridge: Cambridge University Press.

Atwood, D.A. (1990) 'Land Registration in Africa: The Impact on Agricultural Production', *World Development*, 18(5): 659–671.

Baker, R. (1978) 'Barriers to Efficient Capital Investment in Agriculture', in T.W. Schultz (ed.), *Distortions of Agricultural Incentives*, Bloomington: Indiana University Press.

Bardhan, P. (1989) 'Alternative Approaches to the Theory of Institutions in Economic Development', in P. Bardhan (ed.), *The Economic Theory of Agrarian Institutions*, 3–17, Oxford: Clarendon.

—— (1993a) 'Symposium on Management of Local Commons', *The Journal of Economic Perspectives*, 7(4): 87–92.

—— (1993b) 'Analytics of the Institutions of Informal Cooperation in Rural Development', *World Development*, 21(4): 633–639.

Bardhan, P. and Ghatak, M. (1999) *Inequality, Market Imperfections and Collective Action Problems*, Center for International and Development Economic Research, Working Paper No. C99–108, Berkeley: University of California.

Barro, R.J. and Sala-i-Martin, X. (1995) *Economic Growth*, New York: McGraw-Hill.

Becker, G.S. and Barro, R.J. (1988) 'A Reformulation of the Economic Theory of Fertility', *Quarterly Journal of Economics*, 103(1): 1–25.

Becker, G.S., Murphy, K.M. and Tamura, R. (1990) 'Human Capital, Fertility and Economic Growth', *Journal of Political Economy*, 98(5): S12-S37.

Binswanger, H., McIntire, J. and Udry, C. (1989) 'Production Relations in Semi-arid African Agriculture', in P. Bardhan (ed.), *The Economic Theory of Agrarian Institutions*, Oxford: Clarendon.

Binswanger, H., Deininger, K. and Feder, G. (1995) 'Power, Distortions, Revolt and Reform in Agricultural Land Relations', in J. Berhman and T.N Srinivasan (eds), *Handbook of Development Economics*, vol. III, Amsterdam: Elsevier.

Birdsall, N. (1988) 'Economic Approaches to Population Growth and Development', in H.B. Chenery and T.N. Srinivasan (eds), *Handbook of Development Economics*, Amsterdam: Elsevier.

Blaikie, P. and Brookfield, H. (1987) 'Approaches to the Study of Land Degradation', in P. Blaikie and H. Brookfield (eds), *Land Degradation and Society*, London: Methuen.

Bloch, M. (1971) 'The Rise of Dependent Cultivation and Seignorial Institutions', in M.M. Postan (ed.), *The Cambridge Economic History of Europe*, vol. I, Cambridge: Cambridge University Press.

Bongaarts, J. (1995) 'Global and Regional Population Projections to 2025', in N. Islam (ed.), *Population and Food in the Early Twenty-First Century*, Washington DC: IFPRI.

Boserup, E. (1965) *The Conditions of Agricultural Growth*, London: Allen & Unwin.

—— (1981) *Population and Technology*, London: Basil Blackwell.

—— (1989) 'Population, the Status of Women, and Rural Development', in G. McNicoll and M. Caine (eds), *Rural Development and Population* (supplement to vol. 15, *Population and Development Review*), New York: Oxford University Press and The Population Council.

Boyce, J.K. (1987) *Agrarian Impasse in Bengal. Institutional Constraints to Technological Change*, Oxford: Oxford University Press.

Brenner, R. (1976) 'Agrarian Class Structure and Economic Development in Pre-industrial Europe', *Past and Present*, 70: 30–75.

Bromley, D. and Cernea, M. (1989) 'Management of Common Property Natural Resources: Overview of Bank Experience', in R. Meyers (ed.), *Innovation in Resource Management: Proceedings of the Ninth Agricultural Sector Symposium*, Washington DC: World Bank.

Byerlee, D. and Fisher, K. (2000) 'Institutional and Policy Options for Accessing Modern Biotechnology for the Poor', paper presented at the ICABR Conference, Ravello (Italy), 24–28 August.

Caldwell, J.C. (1996) 'The Mechanism of Demographic Change in Historical Perspective', in N. Fobre, *The Economics of the Family*, Cheltenham: Elgar.

Christodoulou, D. (1990) *The Promised Land. Agrarian Reform and Conflict Worldwide*, London: Zed Books.

Clay, D., Guizlo, M. and Wallace, S. (1994) *Population and Land Degradation*, paper No. 14, Department of Agricultural Economics, Michigan State University, East Lansing.

Cleaver, K. (1993) *A Strategy to Develop Agriculture in Sub-Saharan Africa and a Focus for the World Bank*, Washington DC: World Bank.

Coale, A. and Hoover, E. (1958) *Population Growth and Economic Development in Low-Income Countries*, Princeton NJ: Princeton University Press.

Cuffaro, N. (1988) 'Prezzi e Politiche dei Prezzi degli Alimenti nei Paesi in via di Sviluppo', *La Questione Agraria*, 32: 16–50.

—— (1997) 'Population Growth and Agriculture in Poor Countries: a Review of Theoretical Issues and Empirical Evidence', *World Development*, 25(7): 1151–1163.

Cuffaro, N. and Heins, F. (1998) 'Population growth, Food production and Land Degradation in Poor Countries', in J. Restoin (ed.), *Colloque Mondialisation et Géostrategies agroalimentaires*, Montpellier: INRA.

Dale, P. (2000) 'Plants and Biotechnology', paper presented at the workshop *Biotechnology, Environmental Policy and Agriculture*, European Science Foundation, UK Department of International Development and IPGRI, Rome, 29–30 May.

Dasgupta, P. (1993) *An Inquiry into Well-Being and Destitution*, Oxford: Clarendon.

—— (2000) 'Reproductive Externalities and Fertility Behaviour', *European Economic Review*, 44 (4–6): 619–644.

Dasgupta, P. and Ray, D. (1986) 'Inequality as a Determinant of Malnutrition and Unemployment: Theory', *The Economic Journal*, 96(384): 1011–1034.

David, P.A. (1994) 'Why are Institutions the Carriers of History?: Path dependence and the Evolution of Conventions, Organizations and Institutions', *Structural Change and Economic Dynamics*, 5(2): 205–220.

Dawe, D. (1998) 'Reenergizing the Green Revolution in Rice', *American Journal of Agricultural Economics*, 80(5): 969–972.

de Janvry, A. (1973) 'A Socioeconomic Model of Induced Innovations for Argentine Agricultural Development', *Quarterly Journal of Economics*, 87(3): 410–435.

—— (1981) *The Agrarian Question and Reformism in Latin Americas*, Baltimore: Johns Hopkins University Press.

de Janvry, A., Sadoulet, E. and Fafschamps, M. (1989) 'Agrarian Structure, Technological Innovations and the State', in P. Bardhan (ed.), *The Economic Theory of Agrarian Institutions*, Oxford: Clarendon.

De Walt, B., Stonich, S. and Hamilton, S. (1993) 'Honduras: Population, Inequality and Resource Destruction', in C. Jolly and B. Boyle Torrey (eds), *Population and Land Use in Developing Countries*, Washington DC: National Academy Press.

Deininger, K. and Binswanger, H. (1998) *The Evolution of the World Bank's Land Policy*, Washington DC: World Bank.

Demeny, P. (1986) 'Population and the Invisible Hand', *Demography*, 23(4): 473–487.

Demsetz, H. (1967) 'Towards a Theory of Property Rights', *American Economic Review,* 57: 347–359.

Easterlin, R. (1967) 'Effects of Population Growth on the Economic Development of Developing Countries', *The Annals of the American Academy of Political and Social Sciences*, 369: 98–108.

Eicher, C. (1990) 'Building African Scientific Capacity for Agricultural Development', *Agricultural Economics*, 4: 117–143.

El-Ghonemy, M.R. (1990) *The Political Economy of Rural Poverty,* Routledge: London.

Ellis, F. (1993) *Peasant Economics*, Cambridge: Cambridge University Press.

FAO (2000) 'Agriculture: Towards 2015/30', Technical Interim Report, Rome: FAO.

Falcon, W. (2000) 'Globalizing Germplasm: Barriers, Benefits, and Boundaries', paper presented at the XXIV International Conference of Agricultural Economists, Berlin, 13–18 August.

Feder, G. and Feeny, D. (1991) 'Land Tenure and Property Rights: Theory and Implications for Development Policy', *The World Bank Economic Review*, 5(1): 135–153.

Feder, G. and Noronha, R. (1987) 'Land Rights Systems and Agricultural Development in Sub-Saharan Africa', *Research Observer*, 2(2): 143–169.

Forrester, J. (1971) *World Dynamics*, Cambridge MA: Wright-Allen.

Freebairn, D. (1995) 'Did the Green Revolution Concentrate Incomes? A Quantitative Study of Research reports', *World Development*, 23(2): 265–279.

Friedman, J. (1990) *Game Theory with Applications to Economics,* Oxford: Oxford University Press.

Genicot, L. (1971) 'Crisis: From the Middle Ages to Modern Times', in M.M. Postan (ed.), *The Cambridge Economic History of Europe* vol. I, Cambridge: Cambridge University Press.

Goeschl, T. and Swanson, T. (2000) 'Genetic Use Restriction Technology and the Diffusion of Yield Gains to Developing Countries', *Journal of International Development*, 12(8): 1159–1178.

Greenwald, B. and Stiglitz, J.E. (1986) 'Externalities in Economies with Imperfect Information and Incomplete Markets', *Quarterly Journal of Economics*, 101(2): 229–264.

Hardin, G. (1968) 'The Tragedy of the Commons', *Science*, 162: 1243–1248.

Hayami, Y. and Otsuka, K. (1994) 'Beyond the Green Revolution: Agricultural Development Strategy into the New Century', in J.R. Anderson (ed.), *Agricultural Technology: Policy Issues for the International Community*, Wallingford: CAB International and the World Bank.

Hayami, Y. and Ruttan, V. (1985) *Agricultural Development: An International Perspective*, Baltimore: Johns Hopkins University Press.

Herdt, R.W. (1998) 'Reflections on Keeping Asia's Food Basket Full', *American Journal of Agricultural Economics*. 80(5): 969–972.

—— (1999) 'Enclosing the Global Plant Genetic Commons', Occasional paper, The Rockefeller Foundation, 24 May.

Hicks, J. R. (1966) *The Theory of Wages*, London: Macmillan and New York: St Martin's Press.

Hodgson, D. (1988) 'Orthodoxy and Revisionism in America Demography', *Population and Development Review*, 14(4): 541–569.

Hoff, K. (1993) 'Designing Land Policies: An Overview', in K. Hoff, A. Braverman and J. Stiglitz (eds), *The Economics of Rural Organization,* Oxford: Oxford University Press.

Hoff, K. and Stiglitz, J. (1993) 'Imperfect Information and Rural Credit Markets: Puzzles and Policy Perspectives', in K. Hoff, A. Braverman and J. Stiglitz (eds), *The Economics of Rural Organization,* Oxford: Oxford University Press.

Jodha, N. (1987) 'A Case Study in the Degradation of Common Property Resources in India', in P. Blaikie and H. Brookfield (eds), *Land Degradation and Society*, London: Methuen.

Kangasniemi, J. and Reardon, T. (1997) 'Demographic Pressure and the Sustainability of Land Use in Rwanda', in R. Rose, C. Tanner and M. Bellamy (eds), *Issues in Agricultural Competitiveness*, Occasional Paper No. 7, IAAE.

Kelley, A. (1988a) 'Economic Consequences of Population Change in the Third World', *Journal of Economic Literature*, 26(4): 1685–1728.

—— (1988b) 'Population Pressure, Saving and Investment in the Third World: Some Puzzles', *Economic Development and Cultural Change*, 36: 449–464.

Kelley, A. and Schmidt, R.M. (1994) *Population and Income Change: Recent Evidence*, World Bank Discussion Paper 249.

—— (1996) 'Saving, Dependency and Development', *Population Economics* 9(4): 365–386.

Khan, A. (1991) 'Population Growth and Access to Land: An Asian Perspective', in R. Lee *et al.* (eds), *Population, Food and Rural Development*, Oxford: Clarendon.

Komen, J., Cohen, J., Falconi, C. and Salazar, S. (2000) 'Managing Proprietary Technology in Agricultural Research', paper presented at the ICABR Conference, Ravello, Italy, August 24–28.

Kremer, M. (1993) 'Population Growth and Technological Change, One Million B.C. to 1990', *The Quarterly Journal of Economics*, 108(3): 681–716.

Kuznets, S. (1960) *Demographic and Economic Change in Developed Countries*, Princeton, NJ: Princeton University Press.

—— (1967) 'Population and Economic Growth', *Proceedings of the American Philosophical Society*, 111: 170–193.

Lee, R. (1988) 'Induced Population Growth and Induced Technological Progress: Their Interaction in the Accelerating Stage', *Mathematical Population Studies*, I: 265–88.

—— (1997) 'Population Dynamics: Equilibrium, Disequilibrium, and Consequences of Fluctuations', in M. Rosenzweig and O. Stark (eds), *Handbook of Population and Family Economics*, Amsterdam: Elsevier.

Leff, N.H. (1969) 'Dependency Rates and Saving Rates', *American Economic Review* 59(5): 886–896.

Lele, U. and Stone, S. (1989) *Population Pressure, the Environment and Agricultural Intensification: Variations on the Boserup Hypothesis*, MADIA Paper, Washington DC: World Bank.

Lindner, B. (1999) 'Prospects for Public Plant Breeding in a Small Country', paper presented at the ICABR Conference, Tor Vergata, Rome, 17–19 June.

Lipton, M. (1985) *Land Assets and Rural Poverty*, World Bank Staff Working Paper 744, Washington DC.

—— (1990) 'Responses to Rural Population Growth: Malthus and the Moderns', in G. McNicoll and M. Caine (eds), *Rural Development and Population, A Supplement to Vol. 15. Population and Development Review*, New York: Oxford University Press and The Population Council.

—— (1995) 'Land reform as Commenced Business: the Evidence Against Stopping', in A. de Janvry, S. Radwan, E. Sadoulet and E. Thorbecke (eds), *State, Markets and Civil Organisations*, London: ILO and Macmillan.

—— (1997) 'Accelerated Resource Degradation by Agriculture in Developing Countries? The Role of Population Change and Responses to it', in S.A Vosti and T. Reardon (eds), *Sustainability, Growth and Poverty Alleviation*, Baltimore: IFPRI and The Johns Hopkins University Press.

Livi Bacci, M. (1997) *A Concise History of World Population*, London: Blackwell.

Lynam, J.K. and Blaikie, M. (1994) 'Building Effective Agricultural Research Capacity: The African Challenge', in J.R. Anderson (ed.), *Agricultural Technology: Policy Issues for the International Community*, Wallingford: CAB International and the World Bank.

Malthus, T.R. (1826) *An Essay on the Principle of Population. The Sixth Edition* (*1826*) *with Variant Readings from the Second Edition* (*1803*), E.A. Wringley and D. Souden (eds), London: W. Pickering.

Maredia, M., Byerlee, D. and Maredia, K. (1999) 'Investment Strategies for Biotechnology in Emerging Research Systems', paper presented at the ICABR Conference, Tor Vergata, Rome, 17–19 June.

Mason, A. (1987) 'National Saving Rates and Population Growth: a New Model and New Evidence', in D. Gale Johnson and R. Lee (eds), *Population Growth and Economic Development: Issues and Evidence*, Madison: University of Wisconsin Press.

—— (1988) 'Saving, Economic Growth and Demographic Change', *Population and Development Review*, 14(1): 113–144.

Mason, A. and Fry, M. (1982) 'The Variable Rate-of-Growth Effect in the Life-Cycle Model', *Economic Enquiry*, 20(3): 426–442.

Matlon, P. (1987) 'The West African Semiarid Tropics', in J. Mellor, C. Delgado and M. Blaikie (eds), *Accelerating Food Production in Sub-Saharan Africa*, Baltimore: Johns Hopkins University Press.

Matthews, R.C.O. (1986) 'The Economics of Institutions and the Sources of Growth', *The Economic Journal*, 96(384): 903–918.

Meadows, D.L., Randers, J. and Behrens, W.W. (1972) *The Limits to Growth: a Report for the Club of Rome's Project on the Predicament of Mankind*, New York: Universe Books.

Meertens, H.C.C., Fresco, L.O. and Stoop, W.A. (1996) 'Farming Systems Dynamics: Impact of Increasing Population Density and the Availability of Land Resources on Changes in Agricultural Systems. The Case of Sukumaland, Tanzania', *Agriculture, Ecosystems and Environment*, 56: 203–215.

Migot-Adholla, S., Hazell, P., Blarel, B. and Place, F. (1993) 'Indigenous Land Rights Systems in Sub-Saharan Africa: A Constraint on Productivity?', in K. Hoff, A. Braverman and J. Stiglitz (eds), *The Economics of Rural Organization*, Oxford: Oxford University Press.

Moorehead, R. (1989) 'Changes Taking Place in Common Property Resource Management in the Inland Niger Delta of Mali', in F. Berkes (ed.), *Common Property Resources*, London: Belhaven Press.

Myers, G.W. (1991) 'Land and Power: the Impact of the Land Use Act in Southwest Nigeria', *Land Tenure Center Research Papers*, Madison: University Of Wisconsin.

National Research Council (US) (1986) *Population Growth and Economic Development: Policy Questions*, Washington DC: National Academy Press.

Nerlove, M. and Raut, L.K. (1997) 'Growth Models with Endogenous Population: A General Framework', in M. Rosenzweig and O. Stark (eds), *Handbook of Population and Family Economics*, Amsterdam: Elsevier.

North, D.C. (1989) 'Institutions and Economic Growth: An Historical Introduction', *World Development*, 17(9): 1319–1332.

North, D.C. and Thomas, L. (1971) 'The Rise and Fall of the Manorial System: a Theoretical Model', *The Journal of Economic History*, 31: 777–803.

—— (1973) *The Rise of the Western World: A New Economic History*, Cambridge: Cambridge University Press.

Oldeman, L.R., Hakkeling, R.T.A. and Sombroek, W.G. (1991) *World Map of the Status of Human-Induced Soil Degradation. An Explanatory Note*, ISRIC-UNEP.

Ostrom, E. and Gardner, R. (1993) 'Coping with Asymmetries in the Commons: Self-Governing Irrigation Systems Can Work', *Journal of Economic Perspectives*, 7(4): 93–112.

Ostrom E., Gardner, R. and Walker, J. (1994) *Rules, Games and Common-Pool Resources*, Ann Arbor: The University of Michigan Press.

Otsuka, K. (2000) 'Population Pressure, Land Tenure, and Natural Resource Management', paper presented at the XXIV International Conference of Agricultural Economists, Berlin, 13–18 August.

Otsuka, K. and Place, F. (2001) *Land Tenure and Natural Resource Management: A Comparative Study of Agrarian Communities in Asia and Africa*, Baltimore: Johns Hopkins University Press.

Pardley, P.G., Roseboom, J. and Beintema, N.M. (1997) 'Investments in African Agricultural Research', *World Development*, 25(3): 409–423.

Persley, G. (1994) 'Biotechnology's Promise', in J.R. Anderson (ed.), *Agricultural Technology: Policy Issues for the International Community*, Wallingford: CAB International and the World Bank.

Pickering, D. (ed.) (1987) *African Agricultural Research and Technological Development*, Washington DC: Deutsche Stiftung für Internationale Entwicklung and the World Bank.

Pingali, P. (1990) 'Institutional and Environmental Constraints to Agricultural Intensification', in G. McNicoll and M. Caine (eds), *Rural Development and Population, A Supplement to Vol. 15 Population and Development Review*, New York: Oxford University Press and The Population Council.

—— (1997) 'Agriculture–Environment–Poverty Interactions in the Southeast Asia Humid Tropics', in S. Vosti and T. Reardon (eds), *Sustainability, Growth and Poverty Alleviation*, Baltimore: IFPRI and The Johns Hopkins University Press.

Pingali, P. and Binswanger, H. (1987) 'Population Density and Agricultural Intensification: A Study in the Evolution of Technologies in Tropical Agriculture', in D. Gale Johnson and R. Lee (eds), *Population Growth and Economic Development*: Issues and Evidence, Madison: University of Wisconsin Press.

Pingali, P., Bigot, Y. and Binswanger, H. (1987) *Agricultural Mechanization and the Evolution of Farming Systems in Sub-Saharan Africa*, Baltimore: Johns Hopkins University Press.

Pinstrup-Andersen, P., Pandya-Lorch, R. and Rosegrant, M. (1999) *World Food Prospects: Critical Issues for the Early Twenty-First Century*, Food Policy Report, Washington DC: IFPRI.

Platteau, J.P. (1992) *Land Reform and Structural Adjustment in Sub-Saharan Africa*, Controversies and Guidelines, FAO Economic and Social Development Paper, No. 107, Rome: FAO.

—— (1996) 'The Evolutionary Theory of Land Rights as Applied to Sub-Saharan Africa: A Critical Assessment', *Development and Change*, 27(1): 29–85.

Postan, M.M. (1971) 'Medieval Agrarian Society in its Prime: England', in M.M. Postan (ed.), *The Cambridge Economic History of Europe*, vol.1, Cambridge: Cambridge University Press.

Prahladachar, M. (1983) 'Income Distribution Effects of the Green Revolution in India: A Review of Empirical Evidence', *World Development*, 11: 927–944.

Rao, M. (1994) *Whiter India's Environment? Economic Reform and Ecological Refurbishment*, Amherst: University of Massachusetts.

Rausser, G., Scotchmer, S. and Simon, L. (1999) 'Intellectual Property and Market Structure in Agriculture', paper presented at the ICABR Conference, Tor Vergata, Rome, 17–19 June.

Reardon, T. (1997) 'Using Evidence of Household Income Diversification to Inform Study of the Rural Nonfarm Labor Market in Africa', *World Development*, 25(5): 735–747.

Reardon, T. and Vosti, S. (1997) 'Poverty-Environment Links in Rural Areas of Developing Countries', in S. Vosti and T. Reardon (eds), *Sustainability, Growth and Poverty Alleviation*, Baltimore: IFPRI and The Johns Hopkins University Press.

Robinson, W.I. (2000) Centre for Arid Zone Studies, UK, Personal communication.

Robinson, W. and Schutjer, W. (1984) 'Agricultural Development and Demographic Change: A Generalization of the Boserup Model', *Economic Development and Cultural Change*, 32: 355–366.

Romer, D. (1996) *Advanced Macroeconomics*, New York: McGraw-Hill.

Romer, P. (1986) 'Increasing Returns and Long-run Growth', *Journal of Political Economy*, 94(5): 1002–1037.

—— (1990) 'Endogenous Technological Change', *Journal of Political Economy*, 98(5): 71–102.

Rosegrant, M., Ringer, C. and Paisner, M. (2000) 'Agricultural Research, Technology and World Food Markets', paper presented at the XXIV International Conference of Agricultural Economists, Berlin, 13–18 August.

Rukuni, M., Blaikie, M. and Eicher, C. (1997) *Crafting Smallholder-driven Agricultural Research Systems in Southern Africa*, Staff Paper No. 97–49, Department of Agricultural Economics, Michigan State University East Lansing.

Runge, F. (1992) 'Common Property and Collective Action in Economic Development', in D. Bromley (ed.), *Making the Commons Work. Theory, Practice and Policy*, San Francisco: ICS Press.

Ruttan, V. (1997) 'Sustainable Growth in Agricultural Production: Poetry, Policy and Science', in S.A. Vosti and T. Reardon (eds), *Sustainability, Growth and Poverty Alleviation*, Baltimore: IFPRI and The Johns Hopkins University Press.

Sabourian, H. (1990) 'Repeated Games: A Survey', in F. Hahn (ed.), *The Economics of Missing Markets, Information and Games*, Oxford: Clarendon Press 1990.

Sanderson, W. (1994) 'Simulation Models of Demographic, Economic and Environmental Interactions', in W. Lutz (ed.), *Population, Development, Environment*. International Institute for Applied Systems Analysis, Berlin: Springer-Verlag.

Santaniello, V. (2000) 'Biotechnology and Traditional Breeding in Sub-Saharan Africa', paper presented at the workshop *Biotechnology, Environmental Policy and Agriculture*, European Science Foundation, UK Department of International Development and IPGRI, Rome, 29–30 May.

Schotter, A. (1981) *The Economic Theory of Social Institutions*, Cambridge: Cambridge University Press.

—— (1986) 'The Evolution of Rules', in R. Langlois (ed.), *Economics as a Process – Essays in the New Institutional Economics*, Cambridge: Cambridge University Press.

Schultz, T.W. (1973) *Transforming Traditional Agriculture*, New Haven: Yale University Press.

—— 'The Value of Children: An Economic Perspective', *Journal of Political Economy*, 81(2): S2-S13.

Seabright, P. (1993) 'Managing Local Commons: Theoretical Issues in Incentive Design', *The Journal of Economic Perspectives*, 7(4): 113–134.

Sen, A. (1981) *Poverty and Famines: An Essay on Entitlement and Deprivation*, Oxford: Oxford University Press.

—— (1994) 'Population and Reasoned Agency: Food, Fertility and Economic Development', in K. Lindhal-Kiessling and H. Landberg (eds), *Population, Economic Development, and the Environment*, Oxford: Oxford University Press.

Simon, J. (1977) *The Economics of Population Growth*, Princeton NJ: Princeton University Press.

—— (1981) *The Ultimate Resource*, Princeton NJ: Princeton University Press.

Sjaastadt, E. and Bromley, D. (1997) 'Indigenous Land Rights in Sub-Saharan Africa: Appropriation, Security and Investment Demand', *World Development*, 25(4): 549–562.

Solow, R.M. (1956) 'A Contribution to the Theory of Economic Growth', *Quarterly Journal of Economics*, 70: 65–94.

—— (1987) *Growth Theory: An Exposition*, Oxford: Oxford University Press.

Speirs, M. and Olsen, O. (1992) *Indigenous Integrated Farming Systems in the Sahel*, World Bank Technical Paper 179, Washington DC.

Srinivasan, T.N. and Robinson, J.A. (1997) 'Long Term Consequences of Population Growth: Technological Change, Natural Resources and the Environment', in M.R. Rosenzweig and O. Stark (eds), *Handbook of Population and Family Economics*, Amsterdam: Elsevier.

Stiglitz, J. (1989) 'Rational Peasants, Efficient Institutions, and a Theory of Rural Organization: Methodological Remarks for Development Economics', in P. Bardhan (ed.), *The Economic Theory of Agrarian Institutions*, Oxford: Clarendon.

Stokes, S. and Schutjer, W. (1984) 'Access to Land and Fertility in Developing Countries', in W. Schutjer and S. Stokes (eds), *Rural Development and Human Fertility*, New York: Macmillan.

Stoop, W.A. (1987) 'Variations in Soil Properties Along Three Toposequences in Burkina Faso and Implications for the Development of Improved Cropping Systems', *Agriculture, Ecosystems and Environment*, 19: 241–264.

Tamura, R. (1995) 'Human Capital, Fertility and Economic Growth', in M. Tommasi and K. Ierully (eds), *The New Economics of Human Behaviour*, Cambridge: Cambridge University Press.

Tang, Sui Yan (1994) 'Institutions and Performance in Irrigation Systems', in E. Ostrom, R. Gardner and J. Walker, *Rules, Games and Common-Pool Resources*, Ann Arbor: The University of Michigan Press.

Tobler, W., Deichmann, U., Gottsegen, J. and Maloy, K. (1995) *The Global Demography Project*, National Center for Geographic Information and Analysis, Technical Report TR-95–6, Santa Barbara: University of California.

Traxler, G., Falck-Zepeda, J. and Sain, G. (1999) 'Genes, Germplasm and Developing Countries' Access to Genetically Modified Crop Varieties', paper presented at the ICABR Conference, Tor Vergata, Rome, 17–19 June.

Turner II, B.L., Hyden, G. and Kates, R. (1993) *Population Growth and Agricultural Change in Africa*, Gainesville: University Press of Florida.

Tyffen, M. (1995) 'Population Density, Economic Growth and Societies in Transition: Boserup reconsidered in a Kenyan Case Study', *Development and Change*, 26: 31–66.

Udry, C. (1993) 'Credit Markets in Northern Nigeria: Credit as Insurance in a Rural Economy', in K. Hoff, A. Braverman and J. Stiglitz (eds), *The Economics of Rural Organization: Theory, Practice and Policy*, Oxford: Oxford University Press.

United Nations (1999a) *Population Growth, Structure and Distribution. The Concise Report*, New York: UN Department of Economic and Social Affairs, Population Division.

—— (1999b) *World Population Prospects: The 1998 Revision*, New York: UN Department of Economic and Social Affairs, Population Division.

Valdés, B. (1999) *Economic Growth: Theory, Empirics and Policy*, Cheltenham: Edward Elgar.

Vallaeys, G., Silvestre, P., Blackie, M. and Delgado, C. (1987) 'Development and Extension of Agricultural Production Technology', in J. Mellor, C. Delgado and M. Blackie (eds), *Accelerating Food Production in Sub-Saharan Africa*, Baltimore: Johns Hopkins University Press.

Vierich, H.I.D. and Stoop, W.A. (1990) 'Changes in West African Savanna Agriculture in Response to Growing Population and Continuing Low Rainfall', *Agriculture, Ecosystems and Environment*, 31: 115–132.

Watson, D. (1989) 'The Evolution of Appropriate Resource Management Systems', in F. Berkes (ed.), *Common Property Resources*, London: Belhaven Press.

Weir, D.R. (1988) 'Malthus's Theory of Population', in J. Eatwell, M. Milgate and P. Newman (eds), *The New Palgrave*, London: Macmillan.

Williamson, O. (1986) 'The Economics of Governance: Framework and Implications', in R. Langlois (ed.), *Economics as a Process. Essays in the New Institutional Economics*, Cambridge: Cambridge University Press.

Wilson, C.H. (1977) 'The Historical Study of Economic Growth and Decline in Early Modern History', in E.E. Rich and C.H. Wilson (eds), *The Cambridge Economic History of Europe*, vol. V, Cambridge: Cambridge University Press.

World Bank (1984) *World Development Report 1984*, New York: Oxford University Press.

—— (1991) *The Population, Agriculture and Environment Nexus in Sub-Saharan Africa*, Washington DC: World Bank.

—— (1992) *World Development Report 1992*, New York: Oxford University Press.

Zilberman, D., Yarkin, C. and Heiman, A. (1999) 'Knowledge Management and the Economics of Agricultural Biotechnology', paper presented at the ICABR Conference, Tor Vergata, Rome, 17–19 June.

Index

Entries in *italics* refer to book titles.